OECD Green Growth Studies

Green Growth in Hai Phong, Viet Nam

OECD
BETTER POLICIES FOR BETTER LIVES

This work is published under the responsibility of the Secretary-General of the OECD. The opinions expressed and arguments employed herein do not necessarily reflect the official views of OECD member countries.

This document and any map included herein are without prejudice to the status of or sovereignty over any territory, to the delimitation of international frontiers and boundaries and to the name of any territory, city or area.

Please cite this publication as:
OECD (2016), *Green Growth in Hai Phong, Viet Nam*, OECD Green Growth Studies, OECD Publishing, Paris.
http://dx.doi.org/10.1787/9789264260207-en

ISBN 978-92-64-26016-0 (print)
ISBN 978-92-64-26020-7 (PDF)

Series: OECD Green Growth Studies
ISSN 2222-9515 (print)
ISSN 2222-9523 (online)

Revised version, April 2017.
Details of revisions available at: *www.oecd.org/about/publishing/Corrigendum-GreenGrowthHaiPhong.pdf*.

The statistical data for Israel are supplied by and under the responsibility of the relevant Israeli authorities. The use of such data by the OECD is without prejudice to the status of the Golan Heights, East Jerusalem and Israeli settlements in the West Bank under the terms of international law.

Foreword

Urban green growth policies encourage economic development while reducing either negative environmental externalities (for example, air pollution and carbon dioxide emissions that arise from urban activities) or the consumption of natural resources and environmental assets, including water, energy and undeveloped land. This report analyses the economic and environmental performance of Hai Phong, Viet Nam. It assesses its policies and governance practices to promote green growth, and provides recommendations to enhance its green growth potential.

The findings indicate that despite the remarkable economic progress, Hai Phong's urban development is hindered by challenges affecting the realisation of the city's full economic and environmental performance. These challenges include increased greenhouse gas emissions; air and water pollution; urban sprawl; traffic congestion; ineffective recycling and waste management; and high exposure and vulnerability to seasonal floods. The report identifies both internal and external opportunities the city could harness to overcome these challenges and ensure its long-term economic growth.

This publication is the second case study in the **Urban Green Growth in Dynamic Asia** project. The project explores how to promote green growth in fast-growing cities in Asia by examining policies and governance practices that encourage greening and competitiveness in a rapidly expanding economy. It contributes both to the OECD Green Growth Strategy and to the OECD Strategy on Development, as well as to ongoing discussions on the role of cities in tackling the urgent challenges of climate change. Five case studies have been carried out in Bangkok (Thailand), Hai Phong (Viet Nam), Bandung (Indonesia), Iskandar Malaysia (Malaysia) and Cebu (Philippines). They were chosen on the basis of criteria such as population size, speed of growth, economic structure, and the centrality of the city in the national and regional economy. The results of the case studies will culminate in a synthesis report on Urban Green Growth in Dynamic Asia.

The analysis is based on a "focused comparison" strategy of case study research. This entailed asking the same questions in the different city case studies in order to discern similarities and draw general lessons. Although the analysis focuses on Asian cities, the lessons for promoting green growth will also be relevant for other OECD member countries and cities.

The report was prepared by the Directorate for Public Governance and Territorial Development. The directorate's mission is to help governments at all levels design and implement strategic, evidence-based and innovative policies to strengthen public governance, respond effectively to diverse and disruptive economic, social and environmental challenges, and to deliver on governments' commitments to citizens.

The publication benefited from guidance by the OECD Regional Development Policy Committee and its Working Party on Urban Policy, as well as the support of the local team co-ordinated by the Hai Phong Administration. It draws on data provided by the local team and the OECD Metropolitan Database. The analytical approach draws on *Urban Green Growth in Dynamic Asia: A Conceptual Framework* (2014). The expertise and experience of OECD member countries provided relevant international benchmarks and policy examples.

Acknowledgements

This publication was prepared by Tadashi Matsumoto (Project Manager, Urban Green Growth/Knowledge Sharing), Loïc Daudey (Consultant, Green Growth/Knowledge Sharing), Oscar Huerta-Melchor (Policy Analyst), and Aimee Aguilar-Jaber (Policy Analyst, International Transport Forum), under the supervision of William Tompson (Senior Counsellor), directed by Joaquim Oliveira Martins (Head of the Regional Development Policy Division). It benefited from valuable comments and input from numerous colleagues: Karen Maguire, Peter Borkey, Olaf Merk, Martin Abbott, Kwame Boye Frimpong and Christian Pollok. Ulrike Chaplar, Pilar Philip, Vicky Elliott and Gemma Nellies prepared the manuscript for publication.

Our thanks go to the local team that provided support throughout the study, including the preparation of the background report and organisation of the OECD study missions to Hai Phong. The OECD Secretariat would like to thank in particular: Le Van Thanh (Secretary, Hai Phong Party Committee); Nguyen Van Tung (Chairman, People's Committee of Hai Phong City); Le Khac Nam (Vice Chairman, People's Committee of Hai Phong City); Le Thanh Son (Vice-Chairman, People's Committee of Hai Phong); Nguyen Xuan Binh (Vice-Chairman, People's Committee of Hai Phong); Nguyen Van Binh (Vice Chairman, People's Committee of Hai Phong), Nguyen Anh Tuan (Director, Hai Phong City Department of Foreign Affairs (DOFA)), Vu Hoang Cuong (Manager, Hai Phong City DOFA); Le Thi Hanh (Hai Phong City DOFA), Lien Hoang (Vice Manager, Hai Phong City Department of Planning and Investment), Truong Van Thai (Deputy General Director, Port of Hai Phong JSC), To Long (Infrastructure Management Division, Department of Construction), Pham Huu Huy (Hai Phong Economic Zone Authority) and Cao Van Quy (Vice-Director, Hai Phong Water Supply Joint Stock Company). We also thank the members of the Vietnamese central government who participated in the OECD study missions and provided valuable comments on the report.

The publication also benefited from valuable input from the Knowledge-Sharing Platform on Urban Green Growth in Dynamic Asia and the participants of the fourth OECD Knowledge Sharing Workshop on Urban Green Growth in Dynamic Asia (25-27 June 2015). These include the City of Hai Phong Administration, central ministries and agencies, international organisations and experts, local universities and research institutes, sub-national networks and other Asian and OECD cities. The workshop was supported by the OECD Knowledge-Sharing Alliance. Masato Hayashikawa, Chelsea Roberts, John Lloyd, Amanda Gautherin, Charles Victor and Jeanette Duboys provided the organisational support.

The case study's two international peer reviewers from Sweden and Japan, Viktor Allgurén (City of Gothenburg) and Junichi Sono (City of Kitakyushu), joined the OECD study mission to Hai Phong (22-27 June 2015) and provided valuable reports. Richard Carlos Worden (international environmental consultant) also participated in the mission and provided valuable insights and comments on urban resilience and disaster risk management.

The OECD Secretariat would like to acknowledge gratefully the financial support of the City of Hai Phong, the Government of Japan and the *Deutsche Gesellschaft für Internationale Zusammenarbeit* (GIZ) GmbH.

Table of contents

Tables

Figures

Boxes

Acronyms and abbreviations

3Rs	Reduce, reuse and recycle
ASEAN	Association of Southeast Asian Nations
BOD	Biochemical oxygen demand
BRT	Bus rapid transit
CO_2	Carbon dioxide
CTF	Clean Technology Fund
DPI	Department of Planning and Investment
DWT	Deadweight tonnage
EE	Energy efficiency
EZ	Economic Zone
FDI	Foreign direct investment
FUA	Functional urban area
GDP	Gross domestic product
GEF	Global Environment Facility
GHG	Greenhouse gas
GIS	Geographic Information System
GMS	Greater Mekong Subregion
IC	Industrial Cluster
IZ	Industrial Zone
JICA	Japan International Co-operation Agency
KPI	Key performance indicator
kW	Kilowatt
MOT	Ministry of Transport
MSW	Municipal solid waste
MONRE	Ministry of Natural Resources and Environment
NGO	Non-governmental organisation
ODA	Official development assistance
PM	Particulate matter

PPP	Public-private partnership / Purchasing power parity
SEDP	Socio-economic Development Plan
SME	Small and medium-sized enterprise
TEU	Twenty foot equivalent unit.
UNDP	United Nations Development Program
UNESCO	United Nations Educational, Scientific and Cultural Organization
VDB	Vietnam Development Bank
VGGS	Vietnam Green Growth Strategy
VND	Vietnamese dong
WHO	World Health Organization

Executive summary

This report analyses the economic and environmental performance of Hai Phong, Viet Nam. It assesses its policies and governance practices to promote green growth, and provides recommendations to enhance its green growth potential. Urban green growth policies encourage economic development while reducing either negative environmental externalities (for example, air pollution and carbon dioxide emissions that arise from urban activities) or the consumption of natural resources and environmental assets, including water, energy and undeveloped land.

Hai Phong is the third most populous city in Viet Nam, with a population of 1.96 million as of 2015, and occupies a total land area of 1 527 square kilometres. The city is divided into 15 administrative units, of which 7 are categorised as "urban" areas. The development of the city is remarkable, owing partly to the industrial sector and port, which have been central to its growth. Economic growth, as measured by GDP increased annually between 2010 and 2015 at an average rate of 8.91%. Per capita income in 2015 was USD 3 940, which grew at a rate of 20% annually between 2010 and 2015. This economic growth has reached all levels of society and is reflected in the country's significant progress in reducing poverty. Nonetheless, the city's growth has environmental repercussions, including: increased greenhouse gas emissions, air and water pollution, urban sprawl, traffic congestion, waste recycling, and high exposure and vulnerability to seasonal floods.

With its growing economy, Hai Phong now has a vision for positioning itself as a model green port city, and tackling the negative environmental impacts associated with the city's growth. Many opportunities exist for addressing the city's challenges in the areas of land use and transport, renewable energy and energy efficiency in buildings, and water resources and solid waste management. The city's vulnerability to climate change is also an urgent priority.

Key findings and recommendations

Opportunities for action

While Hai Phong's economic zones (EZs) and industrial clusters (ICs) are expected to contribute to its economic development, they also provide untapped opportunities for promoting green growth. Investment in improving productivity and energy efficiency will also help Hai Phong's industries become more competitive. Air quality in the EZs and ICs has decreased, because of outmoded elimination processes in the production system, insufficient investment and infrequent operation of emissions treatment systems. Real-time monitoring stations (air and water) should be installed in the IZs and ICs and the information made public. Expert inspections should be considered, as well as stringent penalties for non-compliance with standards. One possible option would be to designate a few EZs and ICs as "test-cases" and pilot various instruments to encourage green investment.

A crucial factor for Hai Phong's economic development and attracting of foreign direct investment is the city's skilled labour force. Specialised skills training could help the development and management of modern port-related facilities and equipment, as well as support urban green growth. Hai Phong's universities and institutes should be partners in developing such training programmes for the city.

In terms of spatial planning, urban development in Hai Phong should ensure that job centres have adequate housing nearby. Mixed-use development could bring jobs, homes and public spaces in closer together. A clear strategy for land use that allows for high density and reserves land for public services will be needed, as well as effective development control mechanisms. Furthermore, road network improvements in Hai Phong should be accompanied by major investments in public transport, to prevent motorcycle dependency from becoming car dependency.

A systematic approach to waste recycling will create opportunities for growth in Hai Phong. Its urban areas now generate 1.3 kilogrammes of waste per capita, which exceeds the rates in Singapore (0.87kg) and Seoul (0.95kg). Less than 10% of the municipal waste generated is composted, and most recyclable waste is mixed with other refuse and landfilled. The most important step for Hai Phong is to introduce separation of waste at source, and to increase composting and recycling. Waste-to-energy technology also has considerable potential. Since 60% of electricity is consumed by the industrial sector, support for improving energy efficiency to small and medium-sized enterprises should be made a priority. The potential of solar and wind energy should also be explored. In addition, the national government should consider progressively reducing fossil fuel subsidies.

Hai Phong suffers from periodic floods and other natural disasters. Effective preparation for floods in future will require both "hard" and "soft" policy instruments to allow for quick recovery, especially measures that will have multiple benefits. For example, a green space that can be used as a retention pond in case of flooding can also be used as an urban park protecting the local ecosystem and biodiversity.

Green growth in the Port of Hai Phong

The Port of Hai Phong is the largest seaport in northern Viet Nam, and the second-largest in the country after the port of Ho Chi Minh City. It plays a critical role in the local and regional economy. The dramatic increase in cargo in recent years – from 7.7 million tonnes in 2002 to 41.4 million tonnes in 2014 – has contributed to the city's growth. It has also generated environmental and disaster risks that could significantly undermine the entire city's economic and environmental performance if they are not urgently addressed. A modernisation and expansion programme is currently under way at the port, nonetheless the quality of infrastructure and operations in the overall port system must not be neglected. A multimodal corridor investment plan, including railways and waterways, must be designed to reduce congestion and reliance on trucks for hinterland connections. Port operations should be enhanced by addressing the institutional fragmentation of port activities, and by modernising port labour skills through Hai Phong's Viet Nam Maritime University. Meanwhile, air pollution from port activities should be mitigated by introducing an Environmental Ship Index and onshore power supply. Solid waste pollution should be reduced by setting up a "No Special Fee System" in all Vietnamese ports. Finally, a port resilience action plan aligned with one designed for the whole city could help increase the ports' resilience to weather floods and typhoons.

Governance for urban green growth

Hai Phong's green growth will depend to a large extent on the effectiveness of its governance structure. A review of national strategies and policies in various sectors could encourage greater recognition of the role of Hai Phong and neighbouring cities in pursuing green growth. In collaboration with the central government, Hai Phong authorities may wish to invest in a more collaborative approach to regional/metropolitan planning that involves a broad range of stakeholders, including port authorities. Hai Phong could also establish an interdepartmental Green Growth Co-ordination Council. This would provide a decision-making hierarchy and partnership for the implementation of the Local Action Plan on Green Growth. Alternatively, a joint review board for green growth could be formed to review major infrastructure projects in the city and to ensure that they are aligned with green growth objectives and with other projects.

Hai Phong's Local Action Plan on Green Growth may be constrained by the lack of financial resources. Although the city's real revenue has more than doubled in the past ten years, from Vietnamese Dong (VND) 8 462 billion in 2004 to VND 19 293 billion in 2014, the annual budget remains low, at around USD 334 per inhabitant. To finance green growth, Hai Phong, with the support of the central government, should try to expand own-source revenues, including taxes and fees, to pay for green services. Official development assistance (ODA) is also a potentially significant source of funding for urban green growth, which in the past 15 years has chiefly been devoted to the water supply and sanitation, as well as port development. Official development finance for the port should, as far as possible, support a comprehensive development package to expand the port's hinterland connections and improve its environmental performance. Setting up a local investment fund could help channel more international and private financial resources to the city.

Finally, Hai Phong needs to enhance its relations with its citizens. This could include providing access to reliable and timely information and promoting wider public participation in the planning and implementation of green growth strategies. Enshrining citizens' roles and responsibilities in the legal and regulatory framework, and developing guidelines on how to engage with civil society, will be critical for encouraging good governance for urban green growth.

Chapter 1

Economic and environmental performance in Hai Phong

Chapter 1 examines the economic and environmental performance of the Hai Phong metropolitan region.

After the introductory section, the second section, on the socio-economic profile of the metropolitan area, explains how urbanisation has affected economic activity, poverty, social equity and living conditions. It includes an assessment of: i) demographic changes, ii) economic performance and diversity and iii) long-term challenges for growth.

The third section addresses environmental challenges and the potential for green growth. It analyses the environmental performance of the city and suggests opportunities for green growth. It includes an assessment of: i) energy performance, ii) transport and land-use trends, iii) challenges on water supply and wastewater treatment and solid waste management systems, and iv) the risk of floods.

The last section analyses the local government system in Viet Nam and the institutional profile of the metropolitan area.

Main points

- Hai Phong is the third-largest urban area in Viet Nam, with 1.96 million inhabitants, after the two largest cities (Ha Noi and Ho Chi Minh City). **Rural-urban migration in Hai Phong has occurred at a rapid rate**, with the proportion of the urban population increasing from 34% to 47% between 2000 and 2015, although its population dynamics have been increasingly determined by movements from and to other provinces in Viet Nam.

- Hai Phong is located in a key economic area of Viet Nam, the Red River Delta, where important logistical activity and almost 50% of the country's rice production is concentrated. **Hai Phong is an important node in the Eastern Economic Corridor of the Greater Mekong Subregion.** Hai Phong's port provides Ha Noi and the hinterlands of northern Viet Nam as well as southern China with access to goods and services. The most important port in the northern region, it is the second most important in the country. Hai Phong's port region container throughput accounted for 25% of the total container throughput in Viet Nam in 2014.

- Economic growth measured by gross domestic product (GDP) between 2010 and 2015 increased annually at an average of 8.91%. GDP per capita in 2015 was USD 3 940, 2.5 times higher than in 2010, and grew at a rate of 20% annually during the period. Hai Phong has benefited from the fact that more than 60% of the population is of working age. The **industry and service sectors** are becoming increasingly important, to the detriment of agriculture, forestry and aquaculture. Changes in the economic structure have been accompanied by a steady increase in the literacy rate (to 90% in 2015) and in the share of workers with high-level education (12% of total labour force in 2015). Rapidly increasing levels of personal income have helped to **significantly reduce poverty**. While unemployment in Hai Phong remains low, it has been on an upward trend in the last few years.

- Rapid urbanisation and economic growth have increased the need for sustainable management of resources. **Electricity use is expected to increase almost threefold** between 2013 and 2020. The increase in private motorised vehicles has resulted in severe congestion, and transport-related deaths from PM10 emissions are expected to double between 2007 and 2020 if no action is taken. The availability of surface water is threatened by industrial plants and urban areas that discharge untreated wastewater into lakes and rivers. Domestic waste generated in Hai Phong in 2025 is expected to be four times the levels in 2000. Waste management is a first-order priority; industrial waste, in particular, will constitute 60% of total waste in 2025.

- Hai Phong's geographical location leaves the city **exposed to natural hazards such as storms and typhoons**. Climate change is further increasing the vulnerability of the city, and it faces a serious risk of coastal flooding. Floods are a potential health risk. A WHO analysis conducted in Viet Nam shows that climate change is likely to expose the city to a high degree of health hazards.

1.1. Introduction: Hai Phong's key location for national and regional economic activity

Hai Phong is located in the north-eastern coast of Viet Nam, along the Red River Delta (Figure 1.1). The Red River Delta and the Mekong River Delta in the south are the focus of logistics activity in Viet Nam and accommodate the highest population densities (World Bank, 2014). The Red River Delta accounts for nearly 50% of total rice production in Viet Nam (Neumann et al., 2015). As an international gateway port for northern Viet Nam, Hai Phong is poised to become a major transport hub in the region. The city has the second busiest seaport in the country. In 2014, its ports accounted for 25% of total container throughput in the country and have experienced a dramatic increase in cargo volume throughput over the past 15 years (see Chapter 3). With the ports of the Mekong River Delta, which are part of the Ho Chi Minh City port area, Hai Phong and its satellite ports handle 97% of all container volume coming into Viet Nam (Blancas et al., 2014).

Hai Phong is one of the major nodes in the economic corridors of the Greater Mekong Subregion. From the early 1990s, Cambodia, the People's Republic of China (China), the Lao People's Democratic Republic (Lao PDR), Myanmar, Thailand and Viet Nam, began co-operating to increase their commercial ties. Collaboration through various initiatives includes the Greater Mekong Subregion (GMS) Co-operation Programme.[1] A key part of this co-operation has been the development and upgrading of a transport network for the Mekong Subregion. The network has transport links in various "economic corridors" that run along the Mekong River. Hai Phong lies along the Eastern Economic Corridor, the end node of Highway R5, which crosses northern Viet Nam from Kunming in the province of Guanxi in Southern China. The route between Hai Phong and the capital, Ha Noi, is a key segment of this corridor. The majority of goods entering the Hai Phong port area are either destined for Ha Noi or the hinterlands. The Eastern Economic Corridor is part of the free-trade zone between China and the Association of Southeast Asian Nations (ASEAN) and the shortest route from China to the ASEAN region (ADB, 2015).

Figure 1.1. **Map of Hai Phong City by district**

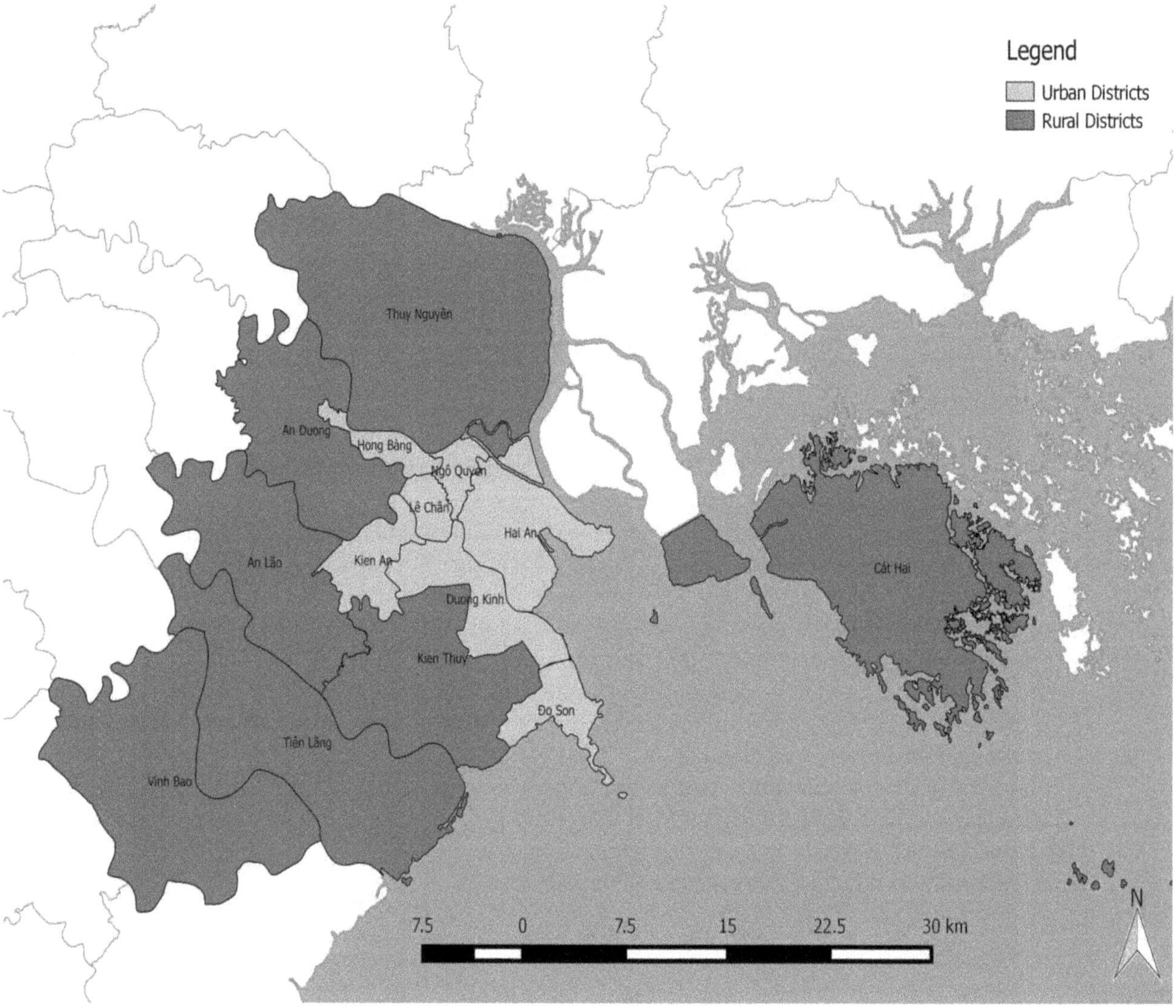

Source: Based on Schneider, A., et al (2015), "A new urban landscape in East-Southeast Asia, 2000-2010", *Environmental Research Letters*, Vol. 10 No. 3, http://iopscience.iop.org/article/10.1088/1748-9326/10/3/034002/meta;jsessionid=9D6988DCBDEE4E3E5D505D371111AE88.c2.iopscience.cld.iop.org.

1.2. Urbanisation and economic transition in Hai Phong

A medium-size urban centre that is rapidly urbanising

With 1.96 million inhabitants, Hai Phong is the third-largest urban area in Viet Nam. However, it is still small in comparison to Ha Noi and Ho Chi Minh City, the two largest urban areas in the country (with populations of 7 million and 9.6 million respectively). Under the national urban classification, Hai Phong is considered as "Grade I" urban areas, while Ha Noi and Ho Chi Minh City are "Special-grade" urban centres (Box 1.1). Hai Phong consists of 15 administrative units, among which seven administrative units are currently recognised as urban districts and make up the so-called "core area". The rest of the administrative units are categorised as "coastal zones" and "ring area" (Do et al., 2012).[2]

Population growth in Hai Phong has been relatively slow. Between 2000 and 2010, the overall population grew from 1.7 million to 1.96 million, the equivalent of a 1% average annual growth. The growth rate is slightly below the trend for the whole Red River Delta, including Ha Noi, and far below that of other major cities, such as Ho Chi Minh City and Da Nang (Figure 1.2).

Despite the modest population growth, **Hai Phong has experienced a rapid increase in the proportion of urban residents**, thanks to intra-city migration from rural to urban parts of the metropolitan area. The area administered by the city has a significantly higher rate of urbanisation than Viet Nam and the Red River Delta as a whole, although rural population still represents a significant proportion, in contrast to Ho Chi Minh City and Da Nang (Figure 1.3). Its urban population increased by 58.4% in the last 15 years (3.9% annually), from 580 000 in 2000 to 918 800 in 2015 (Figure 1.4). The increase in urban population between 2000 and 2010 was accompanied by a decline in the rural population caused by rural-urban migration. From 2010 to 2015, the growth of the urban population slowed. It is worth noting that the rural population also increased marginally in the period. Although the urbanisation rate of Hai Phong was 46.8% in 2015, some of the districts classified as rural are also being urbanised. This has meant a gradual extension of urban space into rural areas, as residents have increasingly preferred to be located on the periphery due to housing affordability. From 2015 to 2025, the urban population is projected to grow at a much faster rate, expanding to 2.4 million (Figure 1.4).

Box 1.1. **Urban population and urban classification in Viet Nam**

There are multiple definitions for urban population in Viet Nam. The United Nations Department of Economic and Social Affairs (UNDESA) (2014) uses a simple definition of "places with 4 000 inhabitants or more". According to the Ministry of Planning and Investment, the urban population is defined as "people living in inner city areas and towns". All residents of other administrative units (i.e. communes) are regarded as rural population. The definition of an area as urban according to the Law on Urban Planning (Decree No. 42/2009/ND-CP on grading of urban centres, dated 7 May 2009), is based on a set of five criteria:

- the functionality should be as a general or specialised centre, to promote socioeconomic development of the whole country or a certain region
- the population should be at least 4 000
- the population density must be appropriate to the size, nature and characteristics of the urban centre grade and must be calculated for the inner area or the township's consolidated street quarter
- the non-agricultural labour force within the inner area or consolidated street quarter must account for at least 65% of total labour
- urban infrastructure facilities should be built in a synchronous manner and fully integrated
- the urban landscape and architecture must comply with approved regulations and promote the well-being of residents.

According to statistics from the Department of Urban Development, Ministry of Construction, by June 2010, the total population of all urban areas was 33.1 million, accounting for 38.6% of the nation's population, of which the inner-city population was 26 million or 30.5% of national population.

Urban areas in Viet Nam are divided into six grades, as follows, according to the Decree 42/2009 ND-CP, and as approved and recognised by the competent state agency:

1. **Special-grade** urban centres include centrally run cities with urban districts, rural districts and satellite urban centres. Viet Nam has two special urban areas: Ha Noi and Ho Chi Minh City.
2. **Grade I** urban areas include centrally run cities with urban districts and rural districts and possibly attached urban centres; and provincial cities with wards and communes. The seven Grade I urban areas include Hai Phong, Da Nang, Can Tho, Hue, Vinh, Da Lat and Nha Trang.
3. **Grade II** urban areas, of which there are 14, consist of provincial cities with wards and communes.
4. **Grade III** urban areas, of which there are 45, include provincial cities or towns with wards and communes.
5. **Grade IV** urban areas consist of provincial towns with wards and communes or district townships with consolidated street quarters.
6. **Grade V** urban areas are district townships with consolidated street quarters and possibly rural residential clusters.

The proportion of the population in each grade of urban area is as follows: Special grade: 9.5%; Grade I: 3.8%; Grade II: 3.7%; Grade III: 4.5%; Grade IV & V: 8.1%.

Source: UNDESA (2014), *World Urbanization Prospects, 2014 revision*, Population Division, http://esa.un.org/unpd/wup/DataSources/ (accessed 1 July 2016); General Statistics Office (2009), "Viet Nam Population and Housing Census 2009: Migration and urbanisation in Viet Nam: Patterns, trends and differentials", Decree No. 42/2009/ND-CP, Ministry of Planning and Investment, http://vietnam.unfpa.org/sites/asiapacific/files/pub-pdf/7_Monograph-Migration-Urbanization.pdf.

Figure 1.2. **Population growth in the Red River Delta and selected cities in Viet Nam**

Index: 1995 = 100

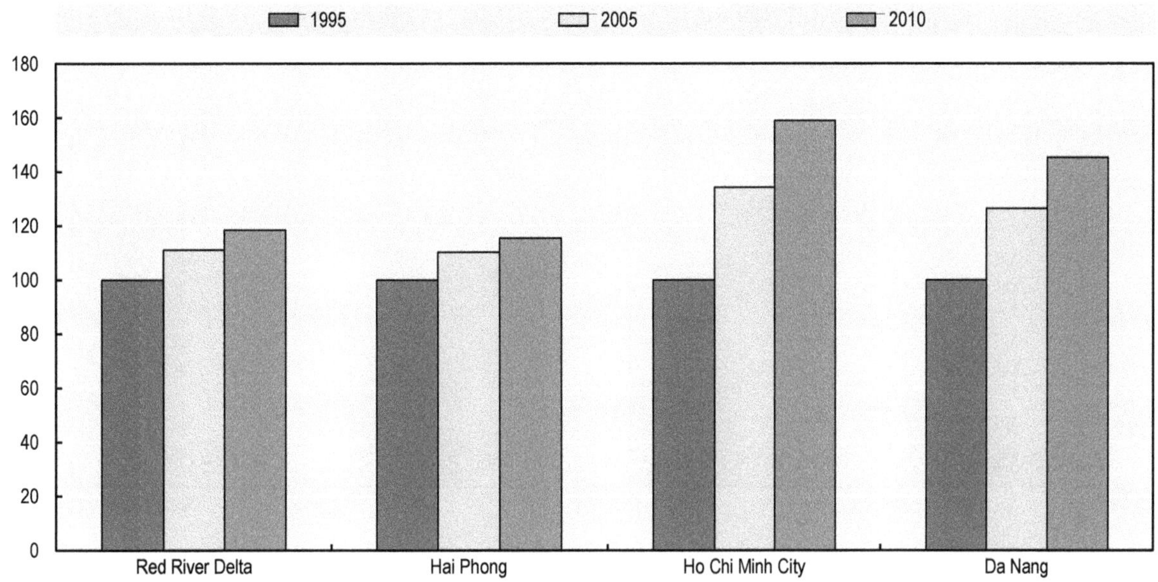

Source: General Statistics Office (2011), "Population projection for Viet Nam 2009-2049", Ministry of Planning and Investment, https://www.gso.gov.vn/default_en.aspx?tabid=617&ItemID=11016.

Figure 1.3. **Urbanisation rate in Viet Nam, the Red River Delta and selected cities in Viet Nam**

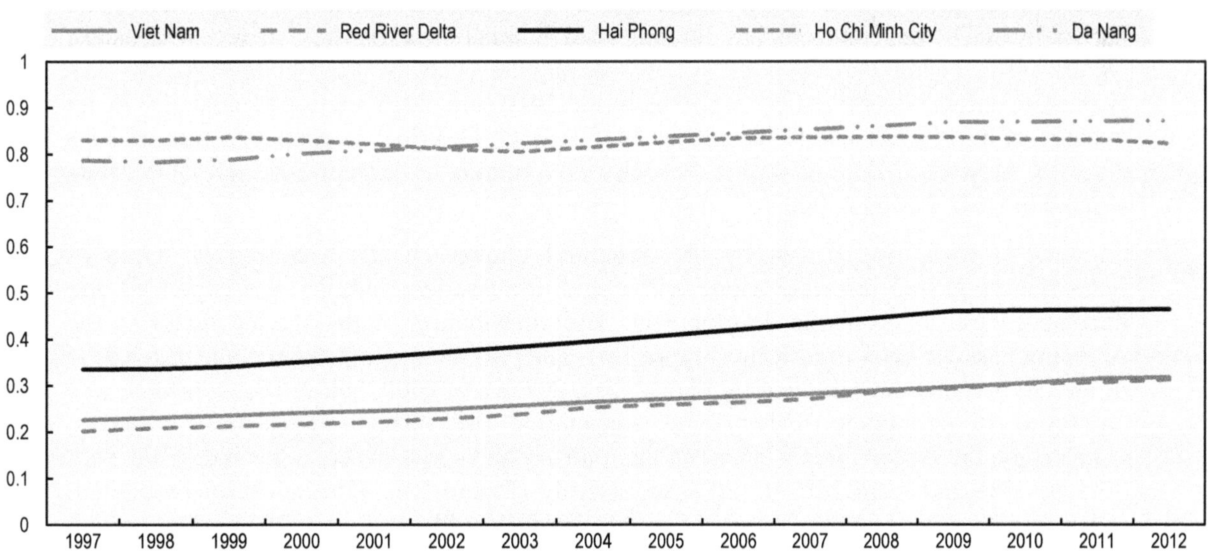

Source: General Statistics Office (2011), "Population projection for Viet Nam 2009-2049", Ministry of Planning and Investment, https://www.gso.gov.vn/default_en.aspx?tabid=617&ItemID=11016.

Figure 1.4. **Urban and rural population in Hai Phong**

2000-15, millions

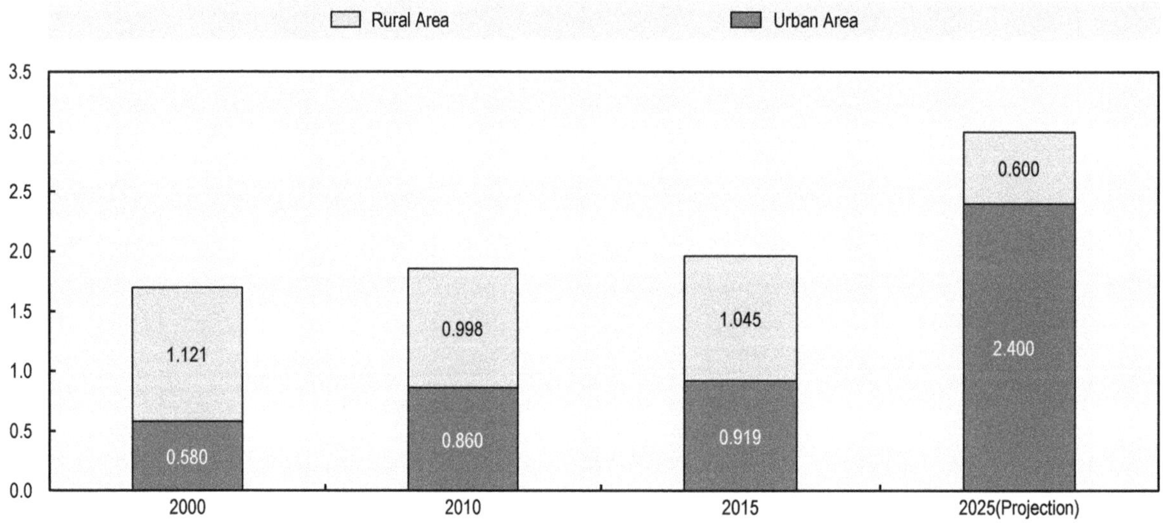

Source: City of Hai Phong (2015a), "Answers to the OECD case study questionnaire", internal document, unpublished.

Rapid economic growth and a changing economic structure

Viet Nam's economy has grown at a rapid pace since the 1980s. In 2013, its real GDP was more than five times what it was in 1980. In the same period, real income per capita increased by a factor of more than eight (Figure 1.5). The economic development of the city is also remarkable, owing to the central role of the industrial sector. Similar to the trend nationally, GDP development has far outpaced population expansion in recent decades, significantly raising the levels of per capita income (City of Hai Phong City, 2015a). Economic growth measured by GDP between 2010 and 2015 increased annually at an average of 8.91%. Per capita income in 2015 was USD 3 941. It grew at a rate of 20% annually between 2010 and 2015. The city government is targeting a GDP per capita income of USD 5 000 in 2020.

GDP growth was driven by the structural change of the economy in terms of industrialisation and services. Along with rapid urbanisation, industry has expanded and become central to economic development. The contribution of the service sector to the economy has also been significant (Figure 1.6). The two sectors in 2014 accounted for 92% of the city's economic activity (Hai Phong City, 2015a). In 2000, half of the workforce was employed in agriculture, fishery and aquaculture, by 2015 only 26% of workers. Conversely, the labour force working in the industry and service sectors has increased from 20% to 27% and from 32% to 47% respectively (Figure 1.7). The industrial production index increased by 16.81% from 2014 to late 2015 (Hai Phong Portal, 2015). The growth was seen across 26 sectors, including manufacturing, processing, production and distribution. The major industries in Hai Phong include shipbuilding and machinery, electronics manufacturing, chemicals, garments, cement and steel. The average growth of exports between 1995 and 2014 was 22% per year (City of Hai Phong, 2015a). Hai Phong attracts large inflows of foreign direct investment (FDI). The total capital investment registered by foreign companies as of early 2016 was over USD 11 billion, representing 450 projects (HEZA, 2016a). In 2015 alone, FDI brought in USD 869 million to the Industrial Zones (IZs) and Enterprise Zones (EZs) in the city.

The favourable conditions for investment and integrated modern infrastructure help account for the increase.

Changes in the economic structure have been accompanied by a continuous increase in literacy rates, to 90% in 2015, and significant growth in the share of workers with a college degree. Nonetheless, this percentage remains low, at 12% compared to Japan (58.6%), Korea (67.7%) and Singapore (28.2%) (OECD, 2016; Government of Singapore, 2016) (Figure 1.8). Hai Phong also has a favourable demographic structure, with over 60% of its population of working age in the period from 2000 to 2015 (Figure 1.9).

Figure 1.5. **Evolution of GDP and GDP per capita in Viet Nam and Hai Phong**

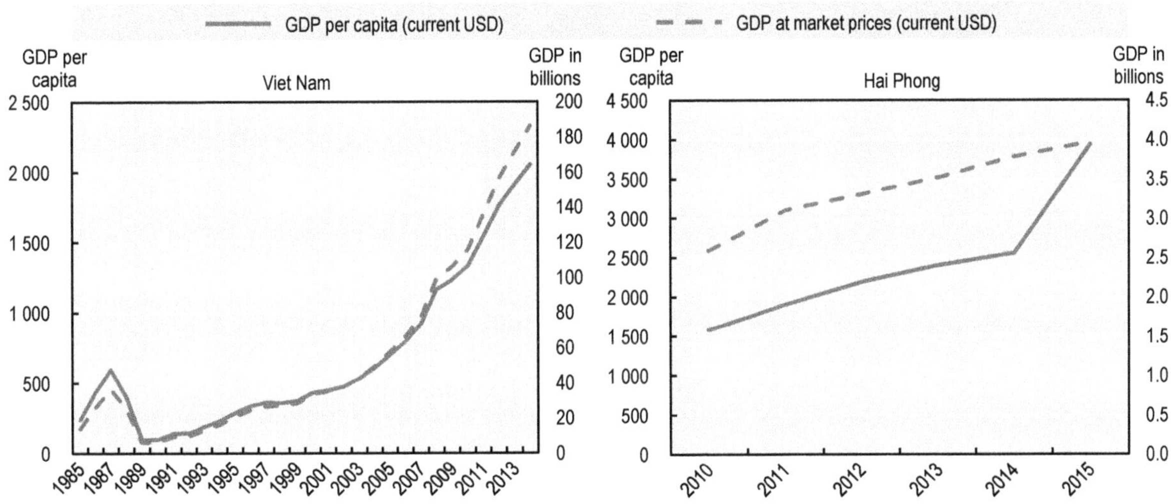

Source: Calculations based on data from the International Monetary Fund (2015), *World Economic Outlook* (database), www.imf.org/external/pubs/ft/weo/2016/01/weodata/download.aspx and General Statistics Office (2015a), *Hai Phong Statistics Yearbook.*

Figure 1.6. **GDP by sector in Viet Nam and Hai Phong**

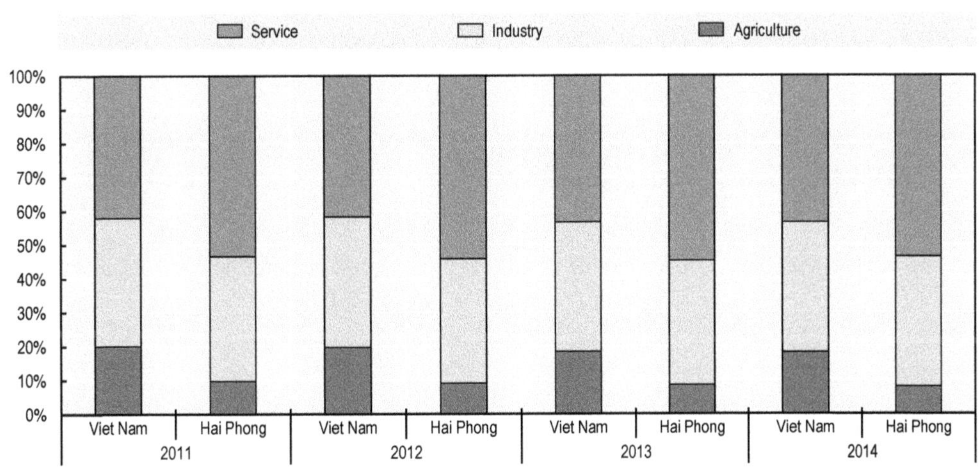

Source: World Bank (2015), *East Asia's Changing Urban Landscape, Measuring a Decade of Spatial Growth,* International Bank for Reconstruction and Development, World Bank, www.worldbank.org/content/dam/Worldbank/Publications/Urban%20Development/EAP_Urban_Expansion_full_report_web.pdf (accessed 1 July 2016) and General Statistics Office (2015a), *Hai Phong Statistics Yearbook.*

Figure 1.7. **Labour force by economic sector in Hai Phong City**

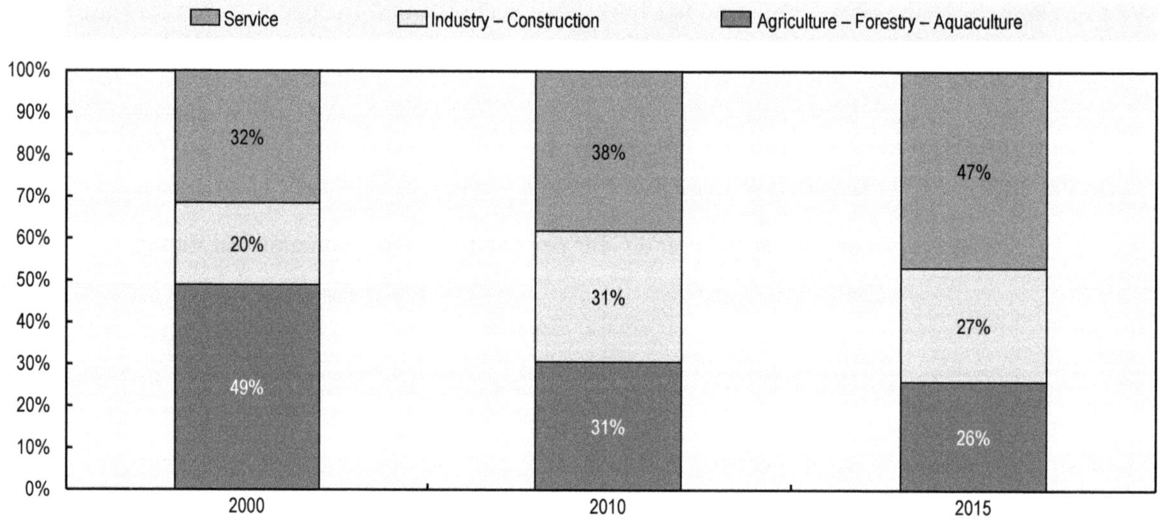

Source: City of Hai Phong (2015a), "Answers to the OECD case study questionnaire", internal document, unpublished.

Figure 1.8. **Education level of labour force**

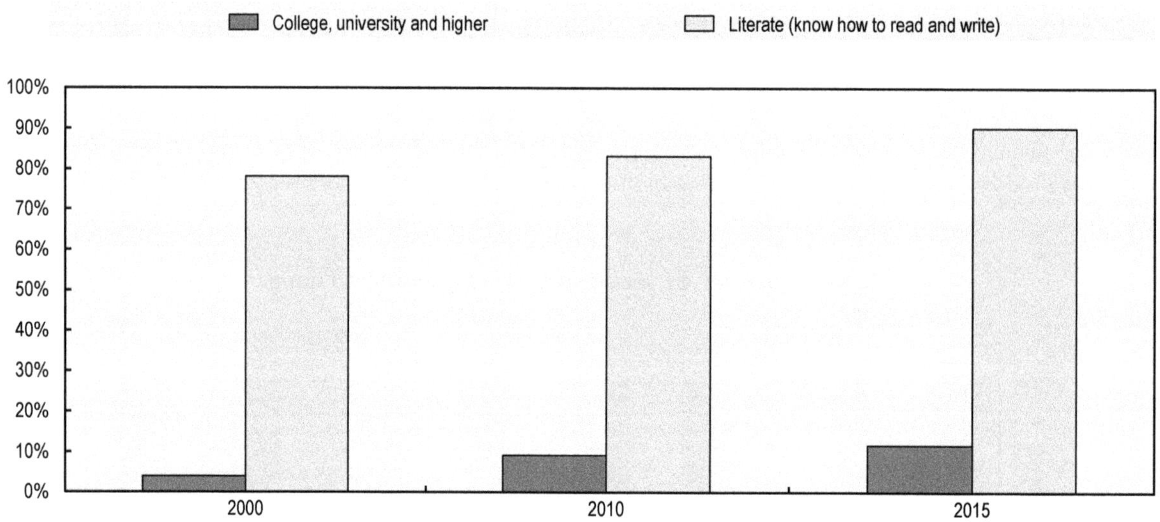

Source: City of Hai Phong (2015a), "Answers to the OECD case study questionnaire", internal document, unpublished.

Figure 1.9. **Population by age cohorts in Hai Phong City**

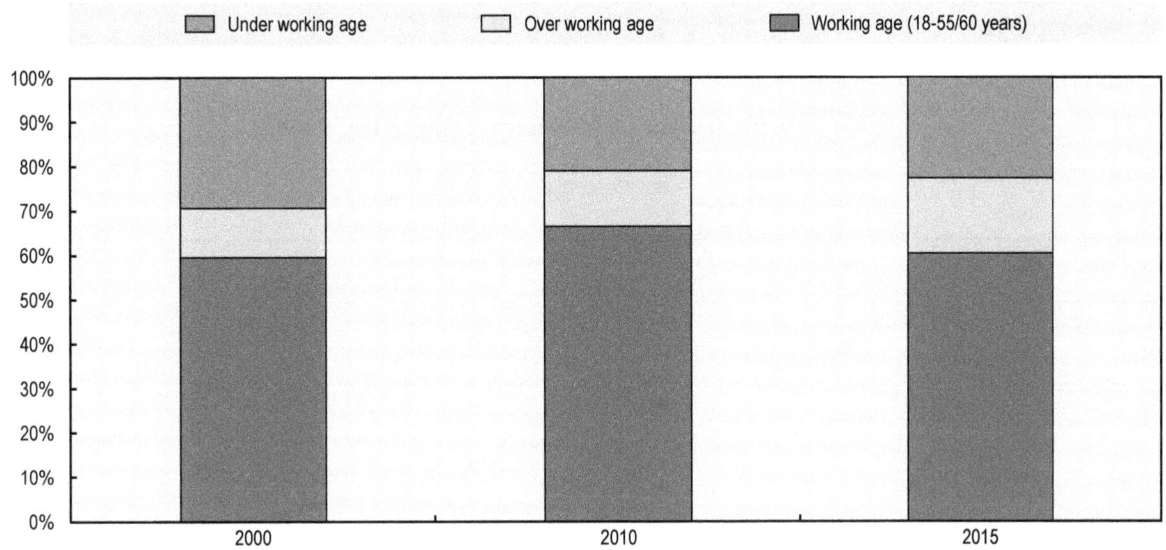

Note: Working age is 18-55 years for women and 18-60 years for men.

Source: City of Hai Phong (2015a), "Answers to the OECD case study questionnaire", internal document, unpublished.

Socioeconomic progress and challenges

The growing levels of personal income in Viet Nam have translated into a rapid reduction in the share of population facing poverty (measured according to income levels). As with the trend nation-wide, Hai Phong has made significant progress in reducing income poverty rates. According to official city data, poverty levels in Hai Phong are decreasing rapidly (Figure 1.10) although data for compounding urban and rural poverty of Hai Phong is not available. The percentage of the poor population in urban and rural areas in Viet Nam implies a significant income gap between the two populations (Figure 1.11). Evaluations of living costs in Hai Phong also indicate that the poverty income threshold should be set at higher levels, at least similar to the poverty line used in Ha Noi, since the cost of living is comparable in the two cities (ActionAid/Oxfam, 2012). The percentage of poor households is determined by the poverty line set by Hai Phong's government.[3] Besides, an analysis conducted in Ha Noi and Ho Chi Minh City suggests a need for taking into account dimensions of poverty other than income levels (Box 1.2).

Data available for Hai Phong also reveal increasing pressure to meet housing demands with solid construction, defined as houses with columns, roofs and walls made of durable materials, such as reinforced concrete, brick, stone, metal or durable wood. About 30% of the new housing stock built between 1999 and 2009 was semi-solid construction, with only two of the main structures (columns, roofs, and walls) made of durable materials. Overall, the share of semi-solid housing increased from 18% to 23% in the period. In addition, while temporary housing (houses where none of the main structures are made of durable materials) remains low, the number of this type of settlement has also been rising (Do et al., 2012).

Hai Phong has also followed the national trend of low unemployment. The unemployment rate in Hai Phong has recently been at 3-4%, all the time lower than the OECD average (Figure 1.12). No data for comparing urban and rural unemployment rates of Hai Phong is available yet. However, analysis of national level data raises the question whether the urban employment market is expanding fast enough to absorb the new urban dwellers. National unemployment data shows that urban unemployment is already higher than rural unemployment albeit declining (Figure 1.13). Although the national urban unemployment rate is falling, the reverse trend taken by the rural unemployment is seen to push up the national rate.

Figure 1.10. **Percentage of poor households in Hai Phong**

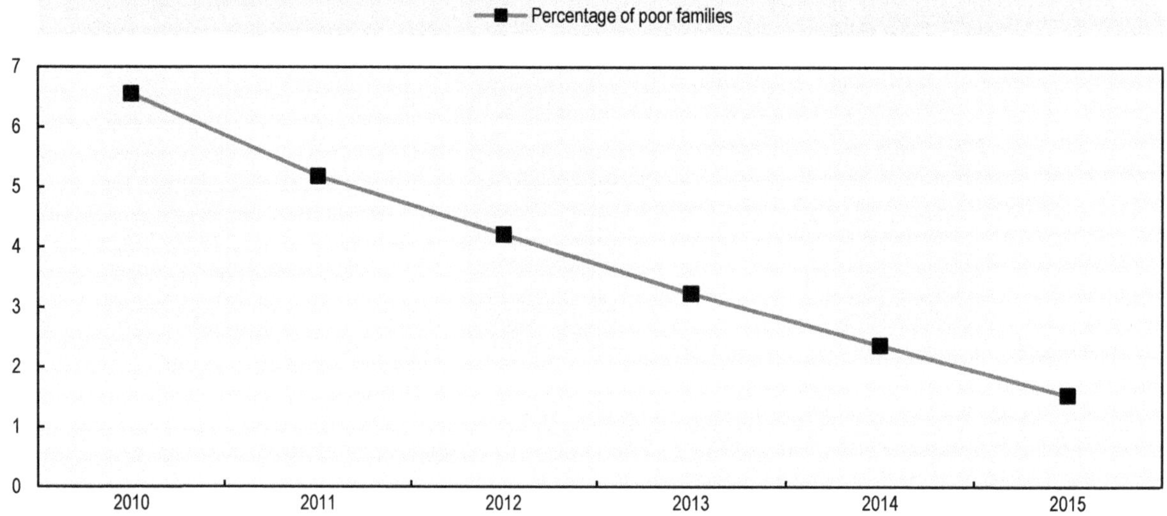

Source: City of Hai Phong (2015a), "Answers to the OECD case study questionnaire", internal document, unpublished.

Figure 1.11. **Share of urban and rural poor in Viet Nam**

Measured according to the General Statistics Office-World Bank methodology

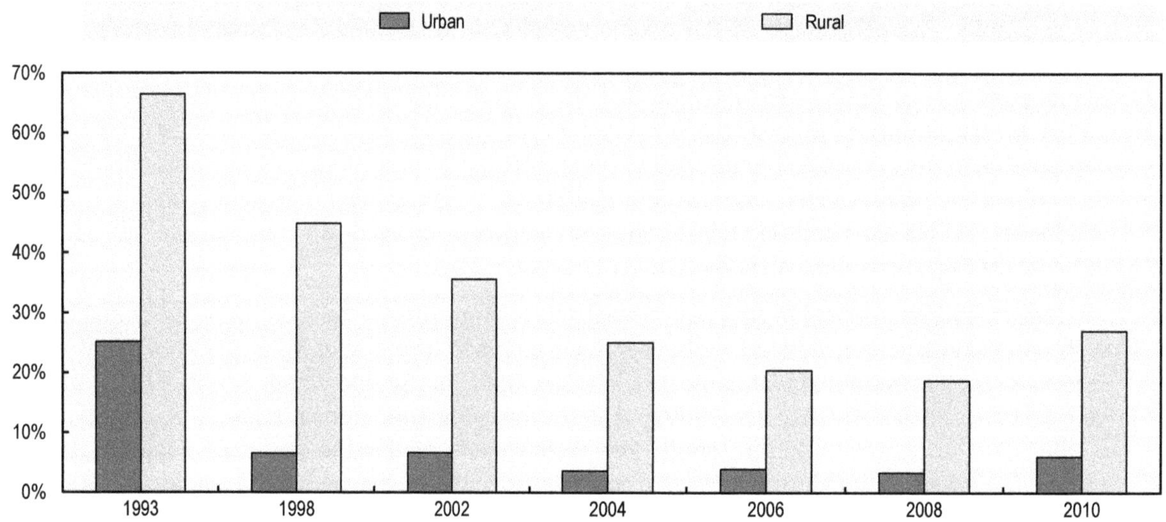

Source: Calculations based on data from the International Institute for Environment and Development (2013), "Urban poverty in Viet Nam – A view from complementary assessments", *Human Settlements Working Paper Series, Poverty Reduction in Urban Areas,* No. 40, http://pubs.iied.org/pdfs/10633IIED.pdf.

Box 1.2. Poverty levels in Ha Noi and Ho Chi Minh City

The United Nations Development Program (UNDP) conducted an urban poverty assessment for Ha Noi and Ho Chi Minh City, to measure different dimensions of poverty. Indicators included several variables in each of eight different dimensions of poverty, including:

1. **Income:** average income per capita per year < VND 6 612 000 (corresponds to USD 2 poverty line).
2. **Education:** schooling age >=18 but without lower-secondary diploma; schooling age from 6 to <18 currently not in school.
3. **Health:** no health insurance due to lack of money or registration, or did not know about health insurance or where to buy it; did not receive health insurance from employers.
4. **Access to social security:** did not receive any benefit from work: severance allowance, delivery/illness benefits, pension, accident insurance, death gratuity; and did not receive pension, regular social allowance; and lived in a household whose members were all of similar status.
5. **Housing quality and area:** housing type is temporary; or roof is made of thatch/oil-paper; or walls are made from dirt/lime/thatch, bamboo wattle/bamboo screen/plywood; or floor is made from clay/earth; without toilet or where the toilet is directly in the water; or area per person is < 7 square metres.
6. **Housing services**: main source of drinking water is not piped or water source is seriously polluted; electricity is not linked to the national gridline or is subject to serious outages, cut-off or voltage variation; garbage is not collected or there is serious pollution from uncollected garbage; no sewage drain.
7. **Social inclusion:** no participation in any social, political organisation, and no participation in any social activities in residential areas.
8. **Physical safety:** living in areas with high or medium levels of theft/robbery or crime.

Results of the assessment show that only 5% and 2% of households in Ha Noi and Ho Chi Minh City are classified as poor when measuring poverty according to income level. However, the proportion of poor households in both cities is significantly higher when assessing other dimensions of poverty. A high percentage of households have little access to social security benefits. In Ha Noi, this percentage is close to 40%, while in Ho Chi Minh City it is above 50% of households. A significant proportion of households in these two cities also have poor housing services (31% in Ha Noi and 36% in Ho Chi Minh City) and or live in houses and housing areas of poor quality (24% in Ha Noi and 29% in Ho Chi Minh City).

Another relevant finding is that in both Ha Noi and Ho Chi Minh City, the percentage of households classified as poor, according to the different dimensions of poverty, is in general higher among migrant households. The difference is significant in dimensions such as access to social security. While 36% of resident households in Ha Noi face poor access to social security, 60% of migrant households face this situation. In Ho Chi Minh City, 22% of resident households have poor housing services. Meanwhile, 67% of migrant households live in such conditions.

Source: UNDP (2010), *Urban poverty assessment in Ha Noi and Ho Chi Minh City*, www.vn.undp.org/content/vietnam/en/home/library/poverty/urban-poverty-assessment-in-ha-noi-and-ho-chi-minh-city.html (accessed 19 July 2016); International Institute for Environment and Development (2013), "Urban poverty in Viet Nam – A view from complementary assessments", *Human Settlements Working Paper Series, Poverty Reduction in Urban Areas,* No. 40, http://pubs.iied.org/pdfs/10633IIED.pdf.

Figure 1.12. **Unemployment rate in Hai Phong City**

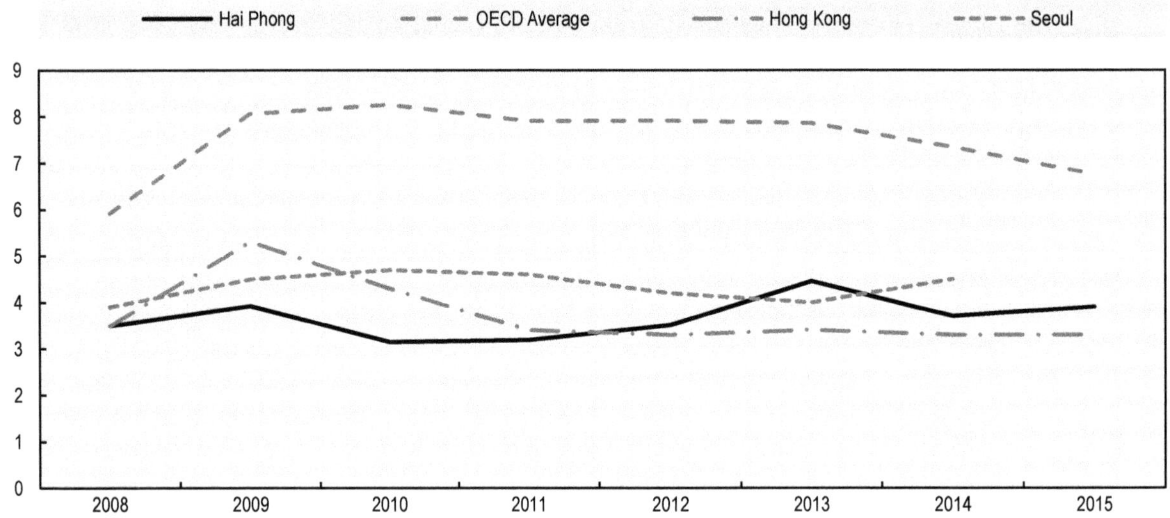

Source: City of Hai Phong (2015a), "Answers to the OECD case study questionnaire", internal document, unpublished; OECD (2014a), "Unemployment rate" (indicator), http://dx.doi.org/10.1787/997c8750-en (accessed on 21 July 2016), Census and Statistics Department (2016), *Labour Force* (database), www.censtatd.gov.hk/hkstat/sub/sp200.jsp?tableID=006&ID=0&productType=8 (accessed 5 July 2016).

Figure 1.13. **Unemployment rates in Viet Nam**

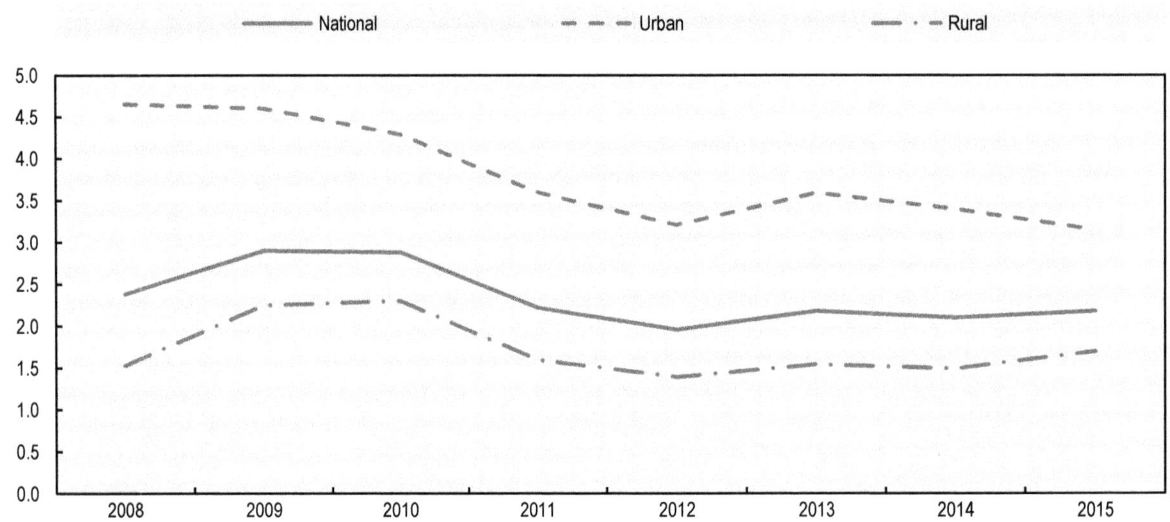

Source: Calculations based on General Statistics Office (2015b), *Unemployment and underemployment rate of labour force in working age by region* (database), www.gso.gov.vn/default_en.aspx?tabid=467&idmid=3&ItemID=12876 (accessed 19 July 2016); City of Hai Phong (2015a), "Answers to the OECD case study questionnaire", internal document, unpublished.

1.3. Environmental trends and green growth challenges in Hai Phong

Growing energy demand in industry and construction sectors

Hai Phong is confronted with a rapidly growing need for more electricity. Under current conditions, it will be unable to meet demand without generating negative environmental and health externalities. By 2020, electric power consumption is expected to rise to almost triple 2013 levels (Figure 1.14). The industry and construction sectors are and will continue to be central to the growth in energy consumption. In 2013, industry and construction accounted for 60% of total electric power consumption. By 2020, the share of electricity consumed by these sectors is expected to grow to 75%. Among the most pressing issues is the high reliance of Hai Phong (and Viet Nam in general) on fossil fuel energy sources, while renewable energy remains marginal. Growth in energy consumption in this context involves rapid growth of GHG emissions and other damaging pollutants harmful to the local environment and public health. The problem is worsening because of limited measures to ensure energy efficiency and conservation in both residential uses and in the growing number of industrial facilities. The important gap in energy consumption relative to GDP between Hai Phong and developed countries in the region emphasises the potential of energy efficiency improvements to help the city cope with the increasing energy demand – Hai Phong consumes 1.14 KWh/GDP, by comparison with 0.19 KWh/GDP in Japan (City of Hai Phong, 2015b).

Figure 1.14. **Energy consumption in Hai Phong in 2013 and 2020, by sector of the economy**

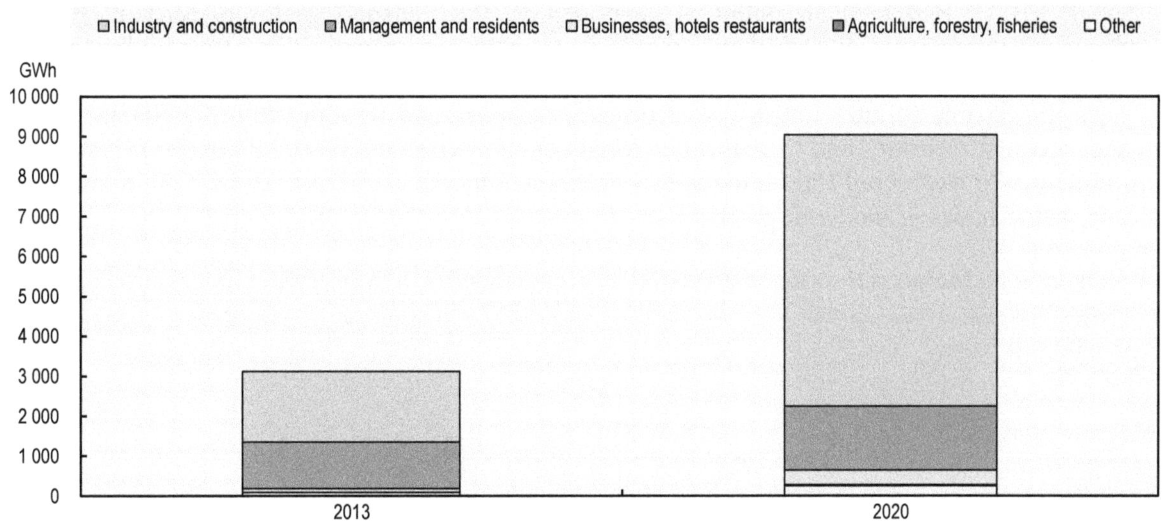

Source: Calculations based on City of Hai Phong (2015b), *Green Growth Promotion Plan of the City of Hai Phong*, http://www.asiangreencamp.net/pdf/green_en.pdf.

Increasing use of motorcycles and low emission standards

Vietnamese transport trends in recent decades have seen extraordinary growth in motorcycle ownership and use. Between 1995 and 2011, the motorcycle fleet in Viet Nam grew at an average yearly rate of 14-15%, from 4 million to 33 million vehicles. In the same period, the number of automobiles grew from 400 000 to almost 2 million vehicles. While the car fleet remains relatively small, the average yearly growth rate for the period ranged between 9% and 10% (UNECE, 2015). The accelerated growth of the private vehicle fleet is especially evident in comparing the relationship between economic growth (GDP/capita) and motorisation rates (vehicles/1 000 population) in Viet Nam and other countries.[4] Viet Nam has much higher motorcycle ownership than other Asian countries of similar per capita income levels (Figure 1.15). In the case of car ownership, while the levels remain low in absolute terms, the intensity of growth relative to income also seems to be higher for Viet Nam than for other countries in the region (Figure 1.16). This national trend also reflects to a great extent the trend in major Vietnamese cities, where bicycles are rapidly being replaced by motorcycles.

The rapid growth in private vehicles, particularly in motorcycles, has become a major source of pollution and GHG emissions in Hai Phong, especially in urban areas. Motorcycles accounted for 66% of total trips realised in 2007 (of a total of 1.94 million trips), while bicycles accounted for 33%. In 2010, the motorcycle motorisation rate reached a level of 363 vehicles per 1 000 population, while the rate for cars was 30 vehicles per 1 000 population (City of Hai Phong, 2015a). In 2014, 78.24% of households owned a motorbike (General Statistics Office, 2015a). The increase in private motorised vehicles has resulted in severe congestion, particularly on National Highway No. 5. While rapid economic growth is probably a major reason for motorisation, poor public transport has also contributed to the trend. Hai Phong has no mass transport infrastructure, and bus services account for a marginal share of total trips (less than 1% in 2007).

In addition to their rapid growth, many of the private vehicles in Hai Phong City are old, and emission standards are lax. If present trends persist, it is estimated that by 2020, annual deaths from mobility-related PM_{10} emissions will be double 2007 levels (Dhondt et al., 2011). Reducing air pollution generated by the transport sector is particularly urgent in urban areas of Hai Phong, where it accounts for around 70% of air pollution (City of Hai Phong, 2015b). This highlights the imperative to i) create a public transport network capable of attracting users of private transport, and ii) adopt transport demand management measures and increase the stringency of environmental standards for private vehicles. For the overall Hai Phong area, a significant modal shift from road freight to rail and water modes is another high priority.

Figure 1.15. **Motorcycle ownership relative to economic growth in selected Asian countries**

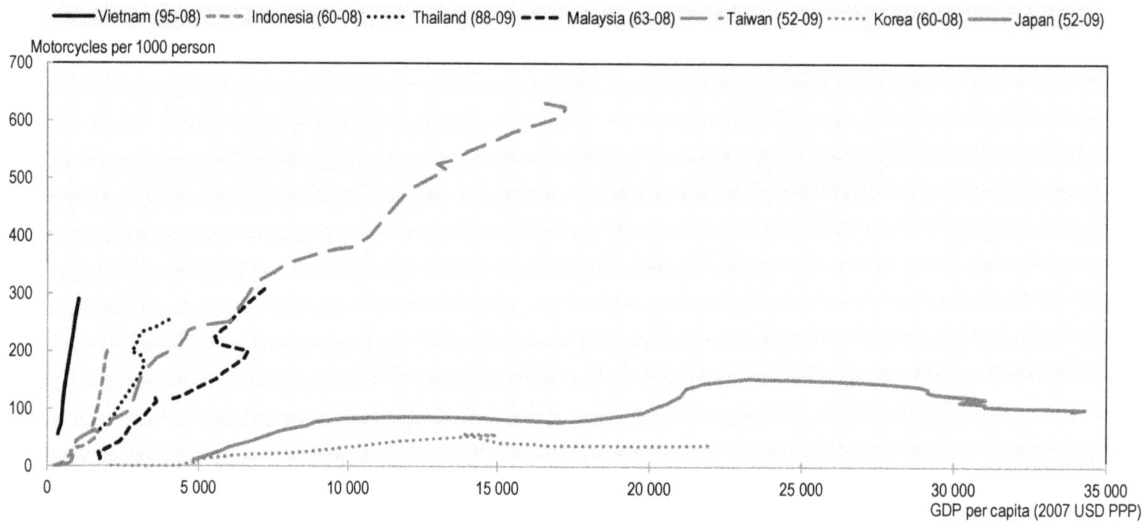

Source: Tuan, V. A. (2011), "Dynamic interactions between private passenger car and motorcycle ownership in Asia: A cross-country analysis", *Journal of the Eastern Asia Society of Transportation Studies*, Vol. 9, https://www.jstage.jst.go.jp/browse/easts/9/0/_contents.

Figure 1.16. **Car ownership relative to economic growth in selected Asian countries**

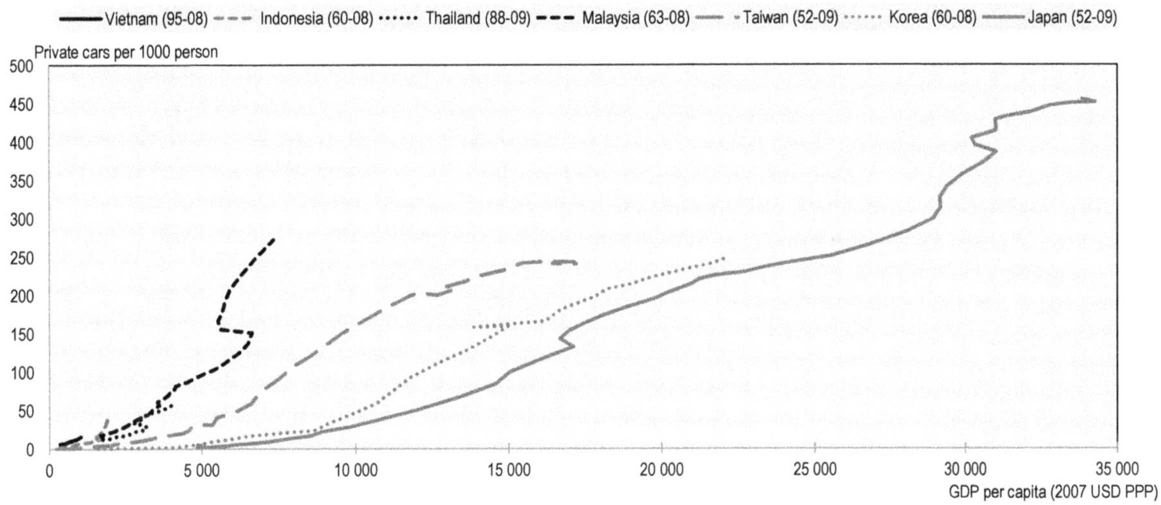

Source: Tuan, V. A. (2011), "Dynamic interactions between private passenger car and motorcycle ownership in Asia: A cross-country analysis", *Journal of the Eastern Asia Society of Transportation Studies*, Vol. 9, https://www.jstage.jst.go.jp/browse/easts/9/0/_contents.

Challenges on water, wastewater and solid waste

Hai Phong, like the rest of Viet Nam, has had difficulty maintaining adequate levels of water supply. Despite its abundant sources of surface water, which account for 2% of the total flow of all rivers globally, Viet Nam experiences higher water scarcity in the dry season from November to April. The Red River basin used to have the capacity to meet water demand in the wet season, as overall demand (including water flows, water stored in reservoirs at the end of the wet season and inter-watershed transfers) did not exceed 20% of supply. In the dry season, the Red River basin had significantly more difficulty keeping up with demand, when demand for water was nearly 40% of the basin's total supply capacity. Such a ratio of demand to supply is considered a medium-level stress zone (ADB, 2009).

This phenomenon can be explained by several factors. First, the impact of climate change may have increased the unequal distribution of rain throughout the year, causing more floods during the rainy season and more severe droughts during the dry season. Secondly, hydroelectric power generation exhausts water flows. In the case of the Red and Thai Binh river basins, 40% of water availability depends on the flows from China (Ministry of Natural Resources and Environment, 2010). In Hai Phong, 27% of the upstream regions of the rivers that reach the city are in Chinese territory. Higher consumption of water from these rivers in China is reportedly reflected in saltwater intrusion that has affected agriculture in Vinh Bao and Tien Lang (City of Hai Phong, 2015b). The problem of decreasing surface water flows has been exacerbated by the exploitation of water resources in major river basins (including the Red and Thai Binh basins), which has exceeded sustainable exploitation levels, which are recommended to be kept below 30% of the total flow (Ministry of Natural Resources and Environment, 2010).

Surface water availability in Hai Phong is also threatened by pollution caused by industrial plants that discharge untreated wastewater into rivers and lakes (Chapter 2). Wastewater from commercial and domestic sources is also released without being treated. In rural districts, sewage release by handicraft villages has further increased water pollution. This situation is exacerbated by the pollution from agriculture, due to the overuse of pesticides and chemicals in farming, which seeps into water sources. Annual fertilizer use is about 24 825 tonnes of urea, 50 428 tonnes of phosphate and 15 130 tonnes of potash. The average annual contents of biochemical oxygen demand (BOD5) and coliform during the 2005-09 period in the Cam River, one of the main sources of surface water for Hai Phong, shows that the water quality has not met the national technical environmental standards (Figure 1.17 and 1.18).

Solid waste in Hai Phong poses another severe challenge. Per capita urban waste generation in 2015 was 475kg per year, which is below the OECD average (516.3kg) but exceeds the average rate for cities such as Kitakyushu (401.9kg) and Singapore (313.9kg) (Figure 1.19). The current waste collected in all the districts of the city is about 2 000 tonnes per day including industrial waste (City of Hai Phong, 2015a). Waste generated in 2025 is expected to be quadruple the levels produced by the city in 2015 (Figure 1.20).

Rapid expansion of the industrial sector will translate into important amounts of industrial waste, which by 2025 will constitute almost 60% of total waste in the city (Figure 1.20). This highlights the need for rapid improvement in managing industrial waste. In particular, adequate management of hazardous residues is of first-order priority. Hazardous waste is often landfilled with general waste, at significant risk to human health and the environment. The amount of hazardous waste generated is equivalent to around 3 288 tonnes. In 2014, about 2 987 tons of hazardous waste was collected and treated. This waste is comprised of sludge from industrial wastewater treatment (28.40%), solution cleaner (27.60%), emulsion and liquid waste without organic halogen compounds (7.59%) and waste oil from ships (6.25%). Other hazardous waste accounted for less than 3%.

Figure 1.17. **Annual average biochemical oxygen demand (BOD5) content in the Cam River, Viet Nam** (mg/L)

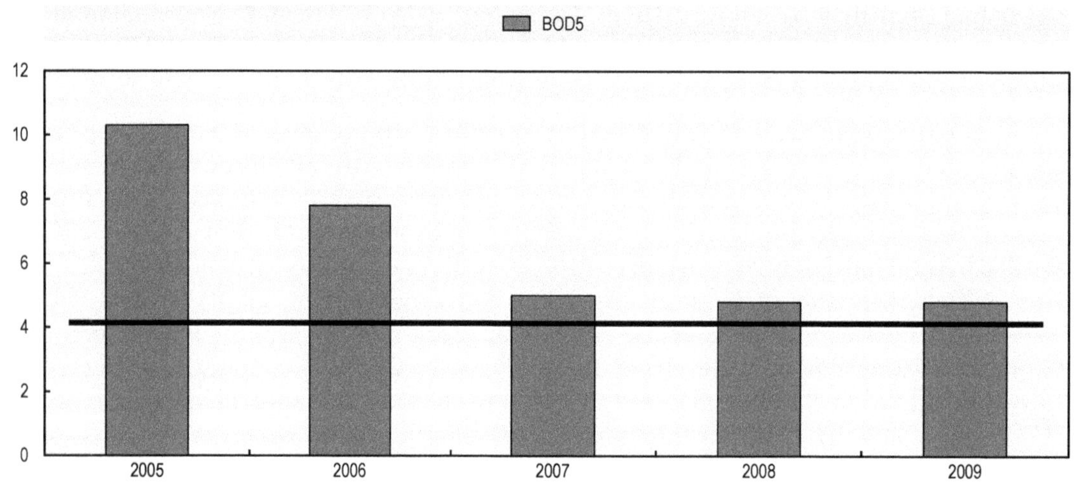

Note: Line denotes the upper limit for domestic water supply purposes according to the National Technical Regulation on Surface Water, issued in 2008 by the Government of the Socialist Republic of Viet Nam.

Source: Ministry of Natural Resources and Environment (2010), *National State of Environment 2010,* Chapter 4, Water Environment, http://quantracmoitruong.gov.vn/portals/0/Chuong%204%20%2893-118%29.pdf?&tabid=142 (accessed 12 July 2016).

Figure 1.18. **Annual average content of coliform in the Cam River, Viet Nam**

(mg/L)

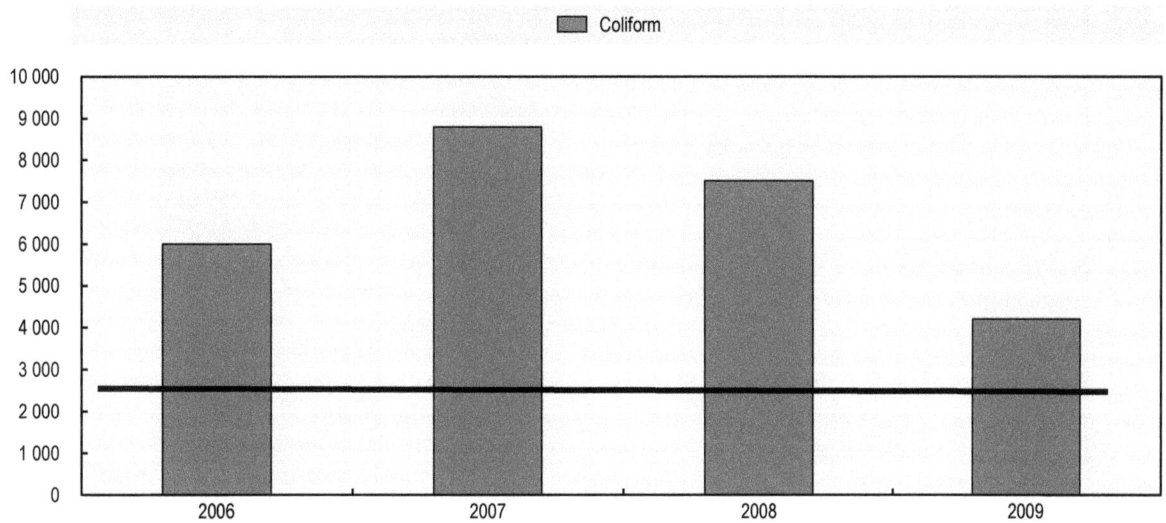

Note: Line denotes the upper limit for domestic water supply purposes according to the National Technical Regulation on Surface Water, issued in 2008 by the Government of the Socialist Republic of Viet Nam.

Source: Ministry of Natural Resources and Environment (2010), *National State of Environment 2010,* Chapter 4, Water Environment, http://quantracmoitruong.gov.vn/portals/0/Chuong%204%20%2893-118%29.pdf?&tabid=142 (accessed 12 July 2016).

Figure 1.19. **Solid waste generation in selected cities**

(kg/capita/year)

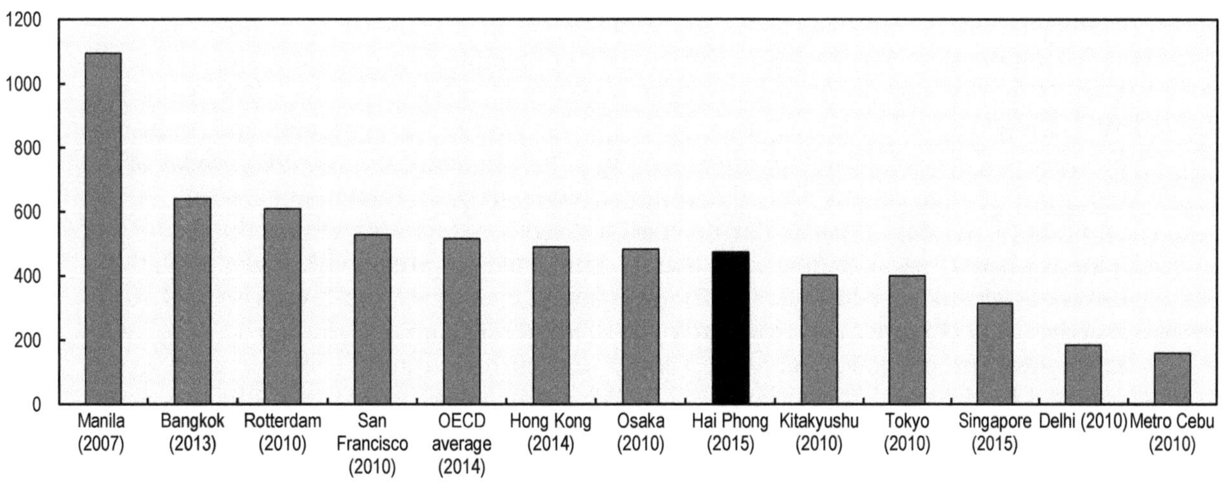

Source: UN-Habitat (2010), *Solid Waste Management in the World's Cities, Water and Sanitation in the World's Cities 2010*; BMA (2014), *Statistical Profile of Bangkok Metropolitan Administration 2013*, Bangkok Metropolitan Administration, Strategy and Evaluation Department, www.bangkok.go.th/main/backoffice/upload_editor/file/stat2013%28ENG%29.pdf; Hoornweg, D. and P. Bhada-Tata (2012), "What a waste: A global review of solid waste management", *Urban Development Series Knowledge Papers*, No. 15, http://siteresources.worldbank.org/INTURBANDEVELOPMENT/Resources/336387-1334852610766/What_a_Waste2012_Final.pdf; Waste Atlas (2016), www.atlas.d-waste.com (accessed 21 June 2016); Social Indicators of Hong Kong (2014), "Per capita municipal solid waste generation", www.socialindicators.org.hk/en/indicators/environmental_quality/23.10 (21 July 2016); JICA (2015), "The Road Map Study for Sustainable Urban Development in Metro Cebu", Final Report.

Figure 1.20. **Waste generated per day in Hai Phong City**

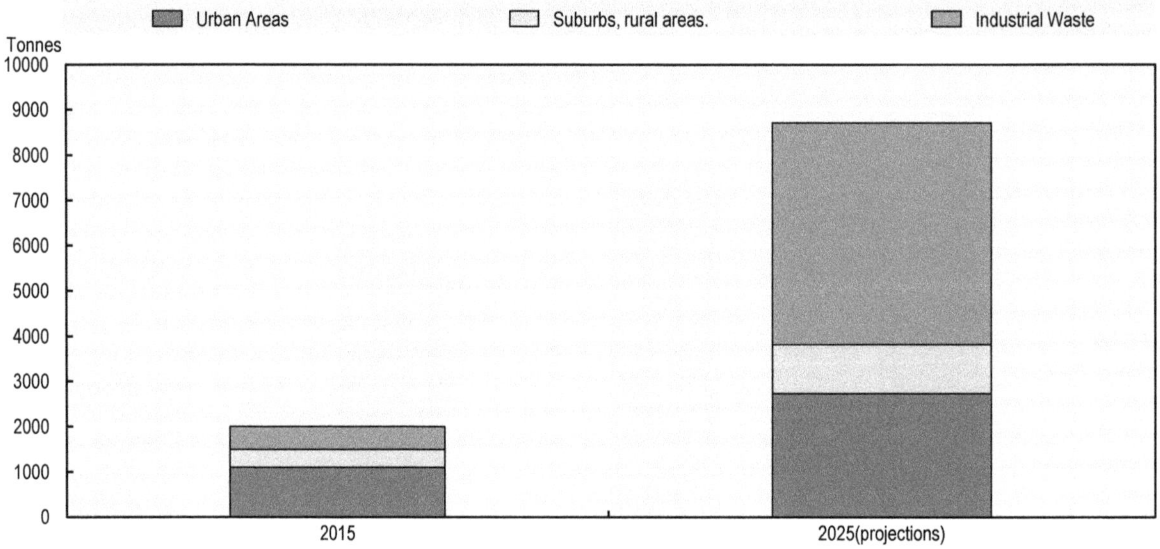

Source: Calculations based on City of Hai Phong (2015b), *Green Growth Promotion Plan of the City of Hai Phong*, http://www.asiangreencamp.net/pdf/green_en.pdf.

A city vulnerable to coastal floods

Viet Nam is subjected to numerous typhoons every year. These natural hazards cause significant damage to the local residential population and infrastructure. Climate change is further increasing the vulnerability of the city, through sea-level rise (Neumann et al., 2015).

Because of its high concentration of people and its coastal location in the Red River Delta region, Hai Phong is particularly vulnerable to natural disasters. In terms of sea-level rise and storm surge, recent assessments show that over 70% of the Red River Delta region's surface area used for residential, commercial and industrial purposes is at risk of a 5-metre high flood by 2050. In addition, over 90 percent of the land used in the region for rice production is at risk (Neumann et al., 2015). Hai Phong City, and in particular its coastal districts, is an area within the Red River Delta at greater risk of being permanently inundated (in areas with lower elevation), as well as at risk of additional temporary floods caused by sea-level rise and storm surge by 2050 (Figure 1.21). Similarly, Hai Phong ranks among the top 15 cities with the highest proportional increase (based on current figures) in exposed assets projected by the year 2070 (Figure 1.22). What this means is that, Hai Phong's assets would be exposed by a factor of about 30 times. Consequently, by that date, under business-as-usual climate change and socioeconomic trends, Hai Phong will be on the top ten list of port cities whose population is most exposed to coastal flooding. Under this scenario, by 2070, Hai Phong is projected to have close to 5 million residents, and over USD 300 billion worth of assets exposed to coastal flooding (Nicholls et al., 2008).[5] It also faces a challenge of accretion and erosion of its coastal fronts and estuaries. Coastal erosion poses a threat in terms of loss of productive land and habitat of coastal organisms (Thanh et al., 2010). Both structural and non-structural measures will be needed to address this natural phenomenon. The accretion has led to a gradual increase of the riverbed, of about 0.2-0.4 metres/year, of rivers such as the Bach Dang River. The uneven distribution of sediment from the accretion is in part accountable for the local erosion in the coastal areas (Duc et al., 2012).

The risks of climate change for Hai Phong include health threats to its population. Evaluation of health vulnerability to climate change has been conducted for provinces in Viet Nam and shows that Hai Phong is less vulnerable to health risks for climate change than other provinces (e.g. Quang Nam, Ninh Binh and Thanh Hoa). Nonetheless, its degree of vulnerability (69/100) is still noteworthy, a result of: i) its geographical and topographic characteristics, since it has major exposure to storms and typhoons; ii) its high population density and the suboptimal living conditions of some of its inhabitants (e.g. children who are not fully vaccinated, population exposed to water pollution, etc.); iii) the limited capacity of its health sector to cope with the levels of the potentially affected population (WHO, 2011).

Figure 1.21. **Areas at risk of sea-level rise and storm surge by 2050**

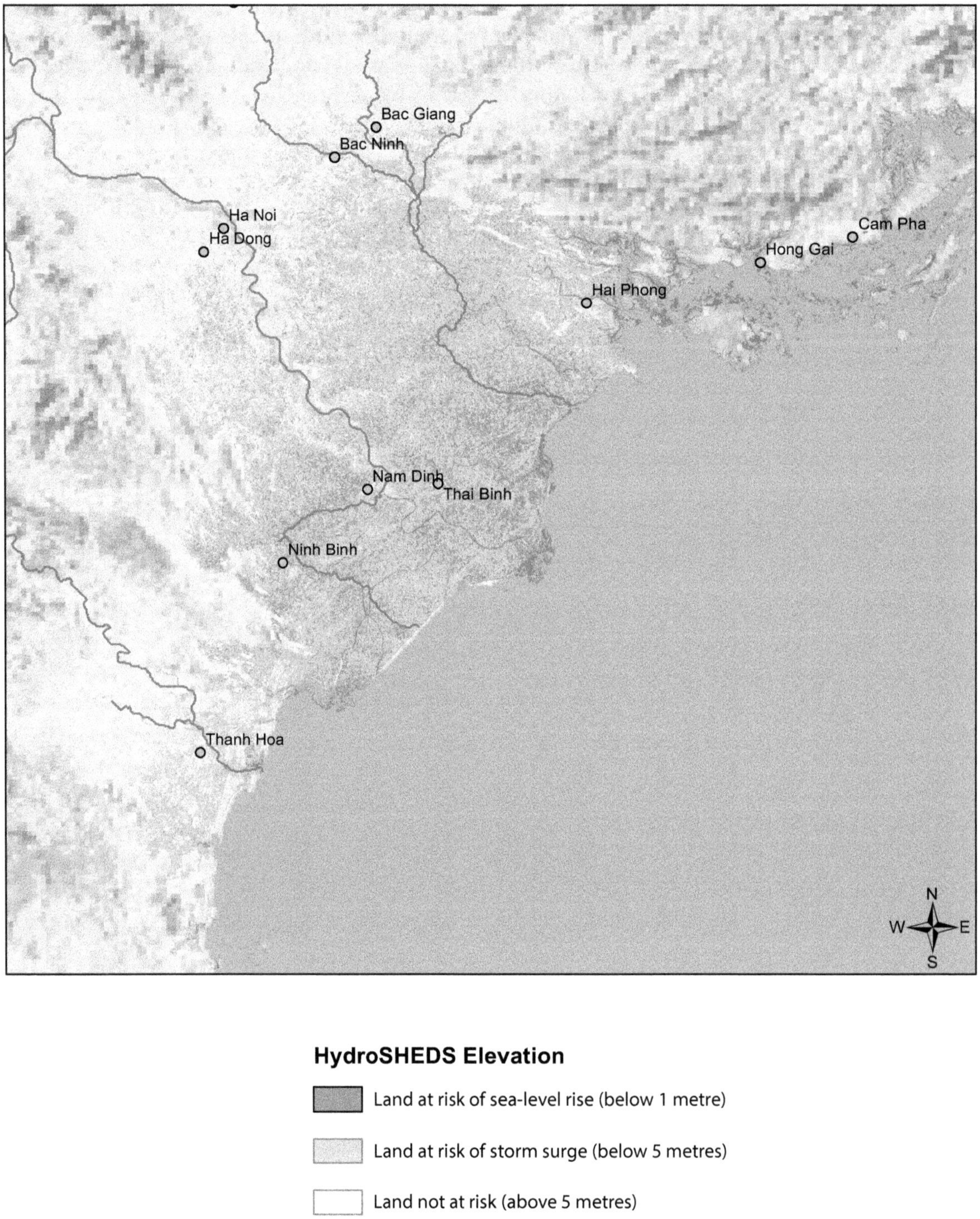

Source: Neumann, J. E. et al. (2015), "Risks of coastal storm surge and the effect of sea level rise in the Red River Delta, Vietnam", *Sustainability*, Vol. 7(6), pp. 6553-6572, http://www.mdpi.com/2071-1050/7/6/6553.

Figure 1.22. **Top 20 cities with the highest proportional increase in exposed assets by the 2070s**

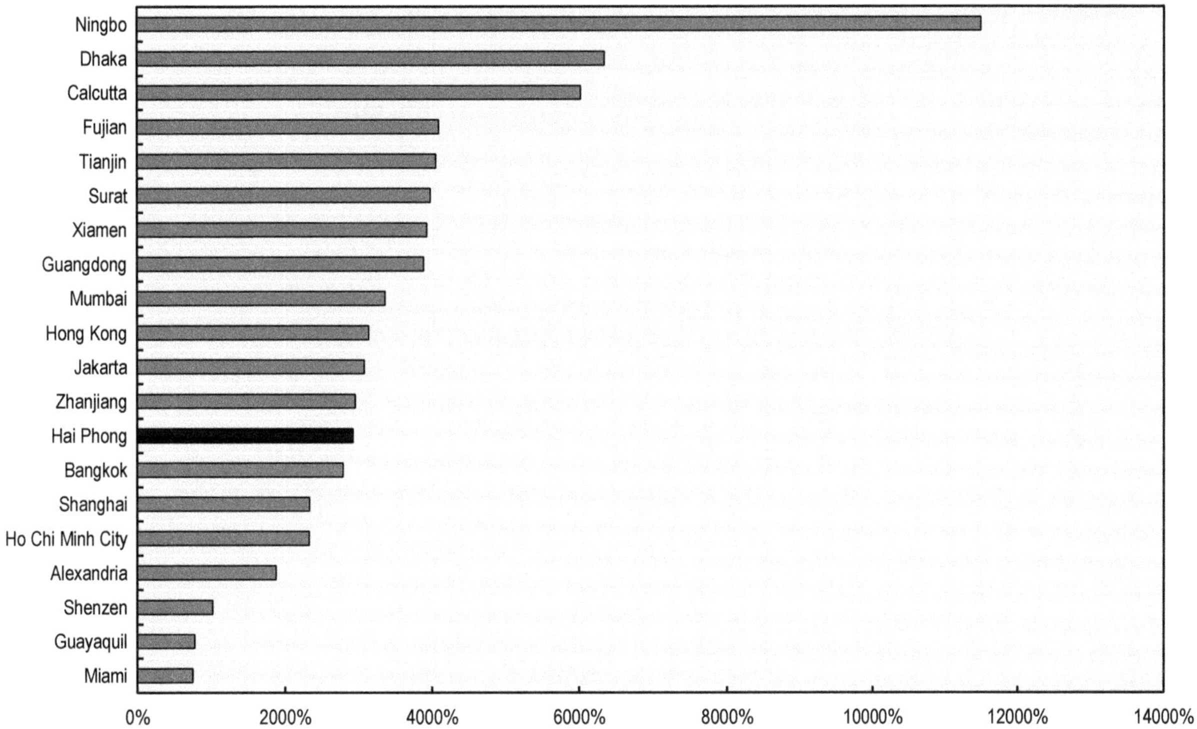

Source: Hanson, S. et al. (2011), "A global ranking of port cities with high exposure to climate extremes", *Climatic Change*, Vol. 104(1), pp. 89-111; IPCC (2012), http://www.lse.ac.uk/CATS/Publications/Publications%20PDFs/83_Ranger_GlobalRanking_2011.pdf; "Summary for policymakers" in *Managing the Risks of Extreme Events and Disasters to Advance Climate Change Adaptation. A Special Report of Working Groups I and II of the Intergovernmental Panel on Climate Change*, pp. 1-19, https://www.ipcc.ch/pdf/special-reports/srex/SREX_Full_Report.pdf.

1.4. Hai Phong's institutional landscape

Hai Phong's city government has the status of a province and is supervised directly by the central government, along with the other four large cities in Viet Nam (Ha Noi, Ho Chi Minh City, Can Tho and Da Nang). Hai Phong is divided into 15 administrative units (7 urban districts, 6 rural districts and 2 island districts) (Figure 1.1).

The local administrative body in Hai Phong, as in all other Vietnamese provincial governments, is the **People's Committee**, headed by a chairman (mayor), which is directed and supervised by a popularly elected local legislative body, the **People's Council**. They are closely related in terms of organisation, function, task and power. The operational effectiveness and efficiency of the local administration largely relies on these two bodies working co-operatively. The members of the People's Committee are appointed by the People's Council. As of 2015, the 223 administrative units were made up of 70 wards, 10 towns and 143 communes under 15 districts (Figure 1.23).

The People's Committee develops the overall budget based on the budgetary plans drawn up by those divisions, and the People's Council discusses and approves the budget. The approved budget is submitted to the level of government directly above it. Under Viet Nam's budget system, the budget must be approved by the People's Council at each level, as well as by the upper level of government. Under the hierarchical fiscal structure, communal governments report to a district government, district governments report to a provincial government and provincial governments report to central government for their respective approvals (Uchimura and Kono, 2012).

Figure 1.23. **Hai Phong's government structure**

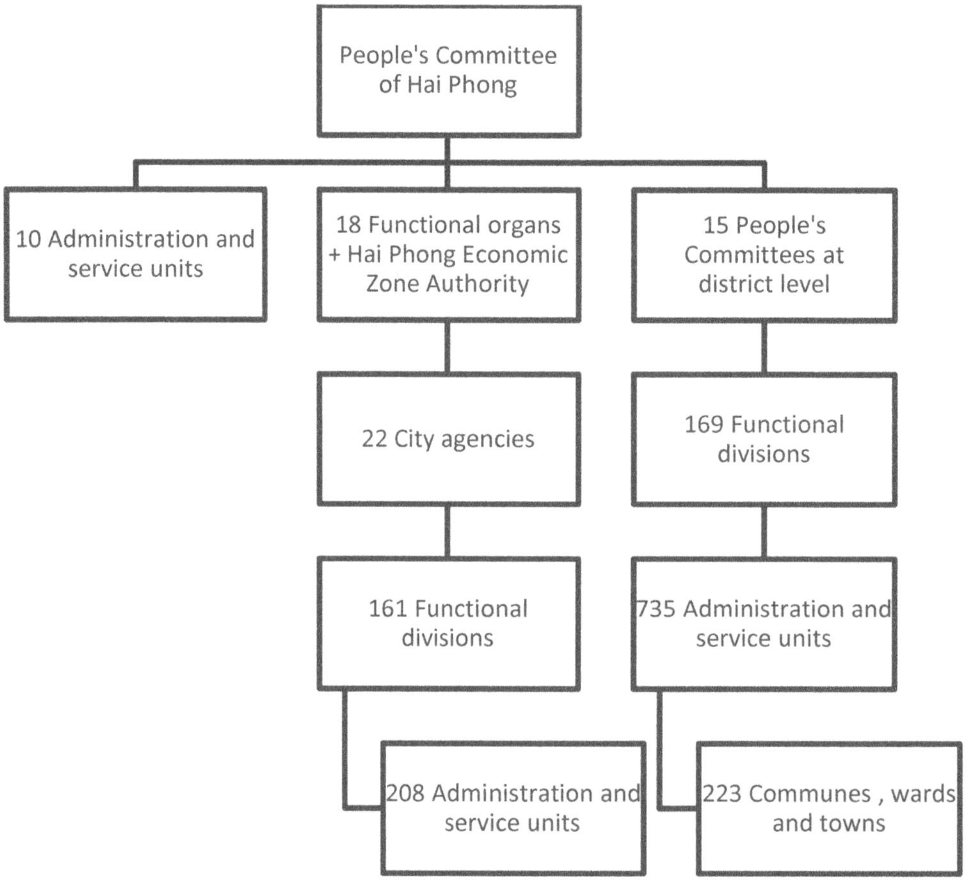

Source: City of Hai Phong (2015a), "Answers to the OECD case study questionnaire", internal document, unpublished.

Notes

1. The programme was launched in 1992 by the six countries in the Greater Mekong Subregion and supported by the Asian Development Bank, to encourage greater economic co-operation. It focuses on promoting co-operation between the countries in the areas of transport, trade facilitation, energy, agriculture, environment, human resource development, urban development, tourism and information and communication technologies (ICT) (ADB, 2015).

2. Hong Bang, Ngo Quyen, Le Chan, Hai An, Kien An, Do Son and Duong Kinh are included in the core area; Thuy Nguyen, Kien Thuy, Tien Lang, and Cat Hai (Bach Long Vi and Cat Ba) are part of the coastal zone; and Vinh Bao, An Duong and An Lao make up the ring area.

3. In Viet Nam, each city and province is permitted to define its own poverty level, as long as it is higher than the levels established by the national government. In Hai Phong, the poverty line is set at a level of VND 400 000 in monthly personal income for rural areas and VND 500 000 for urban areas (ActionAid/Oxfam, 2012)

4. Economic growth is considered a major determining factor in the ownership of private cars. International studies, however, show significant differences between countries and regions in the intensity of the relationship between economic growth and motorisation (Dargay et al., 2007).

5. Assets refer to economic assets such as buildings, transport infrastructure and utility infrastructure (Nicholls et al, 2008).

Bibliography

ActionAid/Oxfam (2012), "Participatory monitoring of urban poverty in Viet Nam, Five-year synthesis report (2008-2012)", http://oxfamblogs.org/Viet Nam/wp-content/uploads/2009/03/Urban-poverty-reduction-report-E.pdf.

ADB (2015), "Greater Mekong Subregion Economic Co-operation Program: Overview", Asian Development Bank, www.adb.org/sites/default/files/publication/29387/gms-ecp-overview-2015.pdf.

ADB (2009), *Socialist Republic of Viet Nam: Water Sector Review*, Asian Development Bank, co-financed by the Government of the Netherlands, Government of Australia and Government of Denmark, http://www.adb.org/sites/default/files/project-document/65077/40621-vie-tacr.pdf .

Blancas, L.C., J. Isbell, M. Isbell, H.J. Tan and W. Tao (2014), *Efficient Logistics, a Key to Viet Nam's Competitiveness*, World Bank, Washington D.C, http://documents.worldbank.org/curated/en/646871468132885170/pdf/830310PUB0978100Box379862B00PUBLIC0.pdf.

BMA (2014), *Statistical Profile of Bangkok Metropolitan Administration 2013*, Bangkok Metropolitan Administration, Strategy and Evaluation Department, Bangkok, www.bangkok.go.th/main/backoffice/upload_editor/file/stat2013%28ENG%29.pdf.

Census and Statistics Department (2016), *Labour Force* (database), www.censtatd.gov.hk/hkstat/sub/sp200.jsp?tableID=006&ID=0&productType=8 (accessed 5 July 2016).

City of Hai Phong (2015a), "Answers to the OECD case study questionnaire", internal document, unpublished.

City of Hai Phong (2015b), *Green Growth Promotion Plan of the City of Hai Phong*, Hai Phong, Viet Nam, http://www.asiangreencamp.net/pdf/green_en.pdf.

Dargay, J., D. Gately and M. Sommer (2007), "Vehicle ownership and income growth, worldwide: 1960-2030", *The Energy Journal*, Vol. 28(4), pp. 143-170.

Dhondt, S., Q. Le Xuan, H. Vu Van and L. Hens (2011), "Environmental health impacts of mobility and transport in Hai Phong, Vietnam", *Stochastic Environmental Research and Risk Assessment: Research Journal*, Vol. 25, pp. 63-376.

Do, T.T.H. et al. (2012), "Human mobility in coastal regions: The impact of migration and temporary mobilities on urbanization", Chapter 10, *Viet Nam: Human Mobility and Urbanization, Archives, Vol.2,* http://digilab-epub.uniroma1.it/index.php/SECOA/article/view/74/63 (accessed 11 July 2016).

Duc, M.D., M.T. Nhuan, and C. Van Ngoi (2012), "An analysis of coastal erosion in the tropical rapid accretion delta of the Red River, Vietnam", *Journal of Asian Earth Sciences*, Vol. 43(1), pp. 98-109, http://www.sciencedirect.com/science/journal/13679120/43.

Hoornweg, D. and P. Bhada-Tata (2012), "What a waste: A global review of solid waste management", *Urban Development Series Knowledge Papers*, No. 15, World Bank, Washington, DC, http://siteresources.worldbank.org/INTURBANDEVELOPMENT/Resources/3363871334852610766/What_a_Waste2012_Final.pdf.

International Institute for Environment and Development (2013), "Urban poverty in Viet Nam", *Human Settlements Working Paper Series, Poverty Reduction in Urban Areas,*No. 40, http://pubs.iied.org/pdfs/10633IIED.pdf.

General Statistics Office (2015a), *Hai Phong Statistics Yearbook,* Ha Noi, Viet Nam.

General Statistics Office (2015b), *Unemployment and underemployment rate of labour force in working age by region* (database), www.gso.gov.vn/default_en.aspx?tabid=467&idmid=3&ItemID=12876 (accessed 19 July 2016).

General Statistics Office (2011), "Population projection for Viet Nam 2009-2049", Ministry of Planning and Investment, Ha Noi, Viet Nam, https://www.gso.gov.vn/default_en.aspx?tabid=617&ItemID=11016 (accessed July 2016).

General Statistics Office (2009), "Viet Nam Population and Housing Census 2009: Migration and urbanisation in Viet Nam: Patterns, trends and differentials", Decree No. 42/2009/ND-CP, Ministry of Planning and Investment, Ha Noi, Viet Nam, http://vietnam.unfpa.org/sites/asiapacific/files/pub-pdf/7_Monograph-Migration-Urbanization.pdf.

Government of Singapore (2016), "Educational Attainment", *Singstat* (database), www.singstat.gov.sg/statistics/visualising-data/charts/educational-attainment (accessed on 28 June 2016).

Hai Phong Portal (2015), "The city's index of industrial growth in 11 months rose 16.55%". http://haiphong.gov.vn/Portal/Detail.aspx?Organization=ubndtp_en&MenuID=3292&ContentID=79782 (accessed, 19 July 2016).

Hanson, S. et al. (2011), "A global ranking of port cities with high exposure to climate extremes", *Climatic Change*, Vol. 104(1), pp. 89-111, http://www.lse.ac.uk/CATS/Publications/Publications%20PDFs/83_Ranger_GlobalRanking_2011.pdf.

HEZA (2016a), *Hai Phong Economic Zone Authority*, www.heza.gov.vn/ (accessed 23 June 2016).

HEZA (2016b), "Dinh Vu Industrial Zone Joint Stock Company Coordinated with Haiphong People's Committee to Hold The Investment Promotion Seminar 2015 In Hanoi", www.heza.gov.vn/Portal/Detail.aspx?Organization=HEZATA&MenuID=4241&ContentID=6099 (accessed 20 July 2016).

International Institute for Environment and Development (2013), "Urban poverty in Viet Nam – A view from complementary assessments", *Human Settlements Working Paper Series, Poverty Reduction in Urban Areas,* No. 40, http://pubs.iied.org/pdfs/10633IIED.pdf.

International Monetary Fund (2015), *World Economic Outlook* (database), www.imf.org/external/pubs/ft/weo/2016/01/weodata/download.aspx.

IPCC (2012), "Summary for policymakers" in *Managing the Risks of Extreme Events and Disasters to Advance Climate Change Adaptation. A Special Report of Working Groups I and II of the Intergovernmental Panel on Climate Change*, Cambridge University Press, Cambridge, UK and New York, pp. 1-19, https://www.ipcc.ch/pdf/special-reports/srex/SREX_Full_Report.pdf.

Mertes, C.M. et al. (2015), "Detecting change in urban areas at continental scales with MODIS data", *Remote Sensing of Environment*, Vol. 158, pp. 331-347, http://dx.doi.org/10.1016/j.rse.2014.09.023.

Ministry of Natural Resources and Environment (2010), *National State of Environment 2010,* Chapter 4, Water Environment, http://quantracmoitruong.gov.vn/portals/0/Chuong%204%20%2893-118%29.pdf?&tabid=142 (accessed 12 July 2016).

Neumann, J.E. et al. (2015), "Risks of coastal storm surge and the effect of sea level rise in the Red River Delta, Viet Nam", United Nations University, World Institute for Development Economics Research, *Sustainability*, Vol. 7(6), pp. 6553-6565, , http://www.mdpi.com/2071-1050/7/6/6553.

Nicholls, R.J. et al. (2008), "Ranking Port Cities with High Exposure and Vulnerability to Climate Extremes: Exposure Estimates", *OECD Environment Working Papers*, No. 1, OECD Publishing, Paris, http://dx.doi.org/10.1787/011766488208.

OECD (2014a), "Unemployment rate" (indicator), http://dx.doi.org/10.1787/997c8750-en (accessed on 21 July 2016).

OECD (2014b), "Population with tertiary education" (indicator), http://dx.doi.org/10.1787/0b8f90e9-en (accessed 21 July 2016).

OECD (2008), *The State of the Public Service*, OECD Publishing, Paris, http://dx.doi.org/10.1787/9789264047990-en.

People's Committee of Hai Phong (2016), "Report of the environmental protection and environmental monitoring results in 2015; tasks and solutions in 2016", People's Committee of Hai Phong, No. 253/BC-UBND, 24 November 2015.

Schneider, A., C.M. Mertes, A.J. Tatem, B. Tan, D. Sulla-Menashe, S.J. Graves, N.N. Patel, J.A. Horton, A.E. Gaughan, J.T. Rollo, I.H. Schelly, F. Stevens and A. Dastur (2014), "A new urban landscape in East-Southeast Asia, 2000-2010", *Environmental Research Letters*, Vol. 10(3), http://documents.worldbank.org/curated/en/844181467988870670/A-new-urban-landscape-in-East-Southeast-Asia-2000-2010.

Social Indicators of Hong Kong (2014), "Per capita municipal solid waste generation", www.socialindicators.org.hk/en/indicators/environmental_quality/23.10 (accessed 21 July 2016).

Thanh T.D, D.V. Huy, N.V. Thao, N.H. Cu, D.L. Land and A.T. Tu (2010), *Coastal Erosion in Cat Hai Island and Solution for Coastal Protection,* www.ccop.or.th/download/delsea/delsea-abstracts.pdf.

Tuan, V. A. (2011), "Dynamic interactions between private passenger car and motorcycle ownership in Asia: A cross-country analysis, *Journal of the Eastern Asia Society of Transportation Studies*, Vol. 9.

Uchimura H. and H. Kono (2012), "Intergovernmental Fiscal Relationships in Vietnam after the 2002 State Budget Law" in *Fiscal Decentralization and Development*, pp.101–132, www.palgraveconnect.com/pc/doifinder/10.1057/9780230389618.0010 (accessed 14 July 2016).

UNDESA (2014) *World Urbanization Prospects, 2014 revision*, Population Division, http://esa.un.org/unpd/wup/DataSources/ (accessed 1 July 2016).

UNDP (2010), *Urban poverty assessment in Ha Noi and Ho Chi Minh City,* www.vn.undp.org/content/vietnam/en/home/library/poverty/urban-poverty-assessment-in-ha-noi-and-ho-chi-minh-city.html (accessed 19 July 2016).

UNECE (2015), *Transport for Sustainable Development; The Case of Inland Transport*, United Nations Publications, New York, www.unece.org/fileadmin/DAM/trans/publications/Transport_for_Sustainable_Development_UNECE_2015.pdf.

UN-Habitat (2010), *Solid Waste Management in the World's Cities, Water and Sanitation in the World's Cities 2010*, UN-Habitat, New York.

Viet Nam Seaports Association (n.d.), www.vpa.org.vn/english/information/info_static2014.html (accessed 23 June 2016).

Waste Atlas (2016), www.atlas.d-waste.com (accessed 21 June 2016).

WHO (2011), *Mapping Out Vulnerable Areas and Population Due to Adverse Health Impacts of Climate Change in Vietnam,* www.wpro.who.int/vietnam/publications/mapping_out_vulnerable_areas_and_community_vietnam_cc_health.pdf.

World Bank (2015), *East Asia's Changing Urban Landscape, Measuring a Decade of Spatial Growth,* International Bank for Reconstruction and Development, World Bank, www.worldbank.org/content/dam/Worldbank/Publications/Urban%20Development/EAP_Urban_Expansion_full_report_web.pdf.

World Bank (2014), "Efficient logistics, a key to Viet Nam's competitiveness", International Bank for Reconstruction and Development, World Bank, Washington DC, http://documents.worldbank.org/curated/en/646871468132885170/pdf/830310PUB0978100Box379862B00PUBLIC0.pdf.

Chapter 2

Hai Phong's opportunities for green growth

Chapter 2 reviews current policies in Hai Phong in the following six areas: energy, land use and transport, housing and buildings, water resource management, solid waste management and green manufacturing. The analysis focuses on the following:

1) Current policies in each area. This will give more precise information on the actions taken by various authorities, and the policy instruments and tools used to reach goals in each of these areas. It helps to identify gaps between existing opportunities and policy responses.

2) Policy outcomes. Where data were available, the impact of each policy was analysed to assess its effectiveness and to identify options for improvement.

3) Policy synergies and complementarities. To strengthen the impact of urban green growth policies, it is important to assess the extent to which Hai Phong has integrated areas of opportunity into coherent and effective policy packages.

The results present several clear opportunities for green growth and call for urgent policy action.

Main points

- The successful growth of Hai Phong's manufacturing industry is remarkable, but it has entailed serious environmental challenges. Hai Phong needs to seek ways to maintain its strong local economic assets while reducing their environmental impact. Opportunities are to be had in the efforts to green local industries. The first step for the city is to improve **pollution monitoring systems** by allocating the necessary resources. In addition, financial support for small and medium-sized enterprises (SMEs) to improve the emission treatment system should be considered urgently. Continuous investment to improve **productivity and energy efficiency in the production system** will create opportunities to make industry more competitive and at the same time more environmentally friendly. Hai Phong has a unique specialisation in port-related industries, and its competitiveness could be enhanced through **specialised job training and skill development** associated with the development and management of modern port-related facilities and equipment, such as information and logistics systems.
- Given Hai Phong's geographical characteristics, **promoting proximity and connectivity of jobs and homes** would help to reduce negative externalities, create economic competitiveness and result in a better quality of life. Redeveloping old and inner port areas will enhance the city's attractiveness and resilience. A **clear land-use vision and effective implementation mechanism** would help the private sector anticipate long-term urban development and to invest in line with this vision. Hai Phong is one of the world's clearest examples of a motorcycle-dependent urban environment. **Prompt and drastic improvements in public transport and adequate demand management for private cars** are needed to head off potential dependency on cars.
- **A comprehensive approach to address water, solid waste and resilience to floods** would help Hai Phong. A starting point is **periodic water quality monitoring**. At the same time, investment in the drainage system and treatment of domestic and commercial wastewater from buildings in the city needs urgent attention. The city is now constructing the first large-scale wastewater treatment plant. While the seven urban districts of Hai Phong city generated 1 037.4 metric tonnes of solid waste per day in 2015, only 100 tonnes were composted, and most recyclable waste was mixed with other refuse and sent to landfill. A systematic approach to promote **recycling** will create opportunities for green growth. The first and most important step is to **separate recyclable goods at the source.** Hai Phong also needs to improve both its "hard" and "soft" infrastructure, as well as its organisational capabilities to more effectively prepare for and recover quickly from future floods. A comprehensive, integrated **floodwater management plan** and adequate climate change **adaptation measures** would help the city address this issue and enhance its resilience.
- Viet Nam's rapidly increasing energy consumption and greenhouse gas (GHG) emissions require urgent action at all levels of government. There are many untapped renewable energy sources the city could explore, such as **waste-to-energy technology** and **wind power generation.** The national government must progressively reduce **fossil fuel subsidies** to divert investment into priority areas and to create a market environment favouring cleaner energy sources. National and local government should work together to promote energy efficient products and buildings, for example through the promotion of **energy labelling** and **green procurement**.

2.1. Promoting greener and more competitive industry

A key green growth opportunity in Hai Phong lies in its efforts to make local industries greener. As discussed in Chapter 1, Hai Phong's strong growth has been driven by structural economic changes, which have shifted from agriculture to manufacturing and service industries. In particular, the manufacturing sector has attracted both domestic and overseas investment, thanks to improved skills and technologies in the labour market. One comparative advantage is the relatively cheap labour force, especially in contrast to People's Republic of China (China). Nearly 70% of FDI projects, both in the number of projects and investment capital, are concentrated in the industrial sector (Table 2.1). The successful growth of manufacturing industries has, however, resulted in serious environmental challenges. Hai Phong needs to seek ways to maintain these strong local economic assets while reducing their negative environmental impact.

Table 2.1. **FDI projects in Hai Phong by sector (2015)**

Investment sectors	Number of projects	Percentage of projects (%)	Investment capital (in USD)	Percentage of investment capital (%)
Industry	288	69.06	6 897 235 900	67.857
Infrastructure – real estate	25	6.00	1 880 915 436	18.505
Hotels, services, education	56	13.43	1 135 883 443	11.175
Transport, trade, agriculture	48	11.51	250 263 419	2.462
Total	417	100.00	10 164 298 198	100.000

Source: City of Hai Phong (2015), "Answers to the OECD case study questionnaire", internal document, unpublished.

Reducing pollution and improving energy efficiency in industry

Hai Phong must take steps to tackle its manufacturing industry, which is highly polluting. Air quality in its industrial zones (IZs) and industrial clusters (ICs) has deteriorated, due to the use of outdated production technologies and limited investment in, and infrequent operation of, emissions treatment systems. Environmental pollution in Hai Phong's IZs and ICs is mainly caused by blast furnace gases (CO, SO_2 and NO_x), dust and noise. For example, steel enterprises in ICs located in the city's west and north-west (Vat Cach – Quan Toan) emit smoke, dust and noise pollution, as well as industrial sewage. Shipbuilding and maintenance, and electricity production, are also responsible for air pollution at some production facilities. Combined, this has had a significant impact on the health of workers and residents in surrounding areas (City of Hai Phong, 2015). The air quality in mining and limestone villages in the suburbs is also alarming, due mainly to drilling for freestone extraction, blasting, crushing and screening stone, as well as transport of the finished stone products. In such villages, as well as those involved in scrap recycling (Trang Minh), mechanical casting (My Dong) and building material processing (Lai Xuan, Anshan), levels of air and noise pollution typically exceed the allowed norms (People's Committee of Hai Phong, 2015). Other villages of copper casting and plastic recycling are polluted by coal combustion and pollutants such as dust, SO_2, CO, NO_x, toxic acids, alkali and metal oxides (e.g. PbO, ZnO and Al_2O_3) (City of Hai Phong, 2015). This not only affects the sustainability of the environment, but has significant ramifications for local public health and well-being.

The first step for Hai Phong is to improve **pollution monitoring systems** by allocating the necessary financial resources for this purpose. The city's monitoring of environmental quality is under-resourced at present. Although the approved Planning Project of environmental monitoring network in Hai Phong (2015-25) has been set up, the installed monitoring sites face financial constraints that have impeded the plan's successful implementation. Consequently, no statistics are available on the contribution of major sources by sector to the city's air pollution. Even though the implementation of the automated environmental monitoring system for major emission units was mandated in the Environmental Protection Regulation and Decree No. 38/2015/ND-CP, no effective monitoring system checks compliance with the national Standard No. 5. The city will need to work with the Department of Industry and the Department of Natural Resources and Environment to make a roadmap and allocate sufficient resources to systematically **install real-time monitoring stations (air and water)** at major IZs, ICs and the port, to identify and measure major pollutants and make the information accessible to public. The self-reporting system and periodical expert inspections should be operated more stringently, by for example increasing penalties for non-compliance. In addition, financial support for SMEs to improve their emission treatment system should be considered a top priority.

Continuous investment to improve **productivity and energy efficiency** will help Hai Phong's industries become more competitive and environmentally friendly. Energy-efficient production systems can cut the cost per unit of production and increase price competitiveness, in addition to consuming less energy. Energy efficiency in industry in the city can also reduce electricity consumption at the national level.

National and local support for the private sector should be scaled up and accelerated. At the national level, the central government has standards, targets and a programme (Master Plan 7) to increase energy efficiency in manufacturing. At the city level, the **Hai Phong Green Growth Strategy for Industry** in 2020 (with the target for 2030) formulates a vision for modernising industry and encourages energy-effective production. Based on this plan, the city will help the 120 casting companies that use coal to transform their manufacturing process. The city is already supporting firms in energy auditing and has provided consulting solutions to save energy. According to the Energy Conservation and Cleaner Production Centre of Hai Phong City (ECC), the energy audit activities have saved a total of 50.3 million kWh per year, equivalent in monetary terms to more than USD 3.03 million (VDN 64 billion) and a reduction in carbon emissions of 42 000 tons per year (ECCH, 2014a). To accelerate these efforts, the city could consider providing **technical guidance and financial support** to all small and medium-sized manufacturing industries in the city. For example, the city could send experts who provide technical advice to these firms and finance a part of the costs of high-energy-efficiency production facilities. The experience of Kitakyushu, Japan, suggests that process conversion and end-of-pipe (EP) technologies can achieve a significant reduction in pollution, and that decisive action by the city was a critical factor in its success (Box 2.1). Iskandar Malaysia's Green Economy Guidelines (GEG) manuals are another promising tool for facilitating green growth in the private sector. The manuals cover a wide range of green actions, such as procurement, operations and supply chain management (Box 2.2).

Box 2.1. **Promoting cleaner production in Kitakyushu, Japan**

In Kitakyushu, a significant reduction in pollution was achieved through cleaner production, including **process conversion** and **end-of-pipe (EP) technologies**. The iron and steel industry further reduced SO_x and other emissions in response to more stringent local pollution monitoring and controls and increasing pressure from the public. The pollution reduction through energy conservation accounted for 33% of the total reduction, the second-largest contribution after the energy supply shift (phasing out coal power plants).

The most effective energy-efficiency measures included replacement of inefficient and polluting equipment with newer and higher-performing technology, e.g. replacing small and middle-size boilers; and optimising production processes by making use of industrial process by-products, such as heat in co-generation. End-of-pipe technology contributed to another 25% of SO_x reduction, in particular desulphurisation treatment and filtering of exhaust gas and gas by-products (EX Corporation, 1996). Some reduction of contaminant levels was also achieved by increasing the height of chimneystacks and releasing pollution higher into the atmosphere. As a consequence of these measures, SO_x was reduced by almost 90% between 1970 and 1990.

Major improvements in water quality were achieved as a result of decisive action to clean up industrial wastewater sediments and undertake large extensions of the public sewerage system. The city faced river and seawater pollution in the 1970s, caused by municipal and industrial wastewater (EX Corporation, 1996). Seawater quality improved after the Dokai Bay clean-up in 1972 and implementation of industrial wastewater regulation.

The city continues to strictly monitor and control local air and water polluters to maintain satisfactory air and water quality levels. Current measures include i) chemical substance monitoring (SO_2, NO_2, CO, SPM, NO_x, $PM_{2.5}$, dioxins, benzene and trichloroethylene) at 14 general ambient air-monitoring stations and 5 automobile exhaust gas monitoring stations, and ii) water quality monitoring at environmental reference points (27 river sites, 7 ocean sites and 1 lake site) and general measurement points (5 river sites, 11 ocean sites). National standards limit the legal amount of air pollution and wastewater discharges from industries and businesses in Kitakyushu, and the city has the power to control the implementation of these standards. In case of non-compliance, the city can require technical improvements, provide guidance to meet regulations or request a temporary shutdown of operations.

High energy efficiency is a key asset for Kitakyushu's heavy industry, which has increasingly focused on resource-efficient production of a number of products. Growing exports of these products ensure the city's economic growth.

Source: EX Corporation (1996), "Kitakyushu - a case study", *Metropolitan Environmental Improvement Program (MEIP)*, World Bank, http://documents.worldbank.org/curated/en/1996/04/441745/japans-experience-urban-environmental-management-vol-2-4-kitakyushu-case-study (accessed 20 July 2016); OECD (2013), *Green Growth in Kitakyushu, Japan*, http://dx.doi.org/10.1787/9789264195134-en.

Box 2.2. **Green Economy Guidelines Manuals in Iskandar Malaysia**

The Green Economy Guidelines (GEG) manuals, published in 2014 by Iskandar Regional Development Authority (IRDA), Malaysia, aim to help business and industry to study, evaluate, adopt and inculcate environmentally sustainable economic behaviour. The aim is to lead to a prosperous, resilient, robust and globally competitive green economy in the Iskandar Malaysia region. They provide a checklist for businesses to address areas of procurement, operations and supply chain management, to minimise impact on the environment (IRDA, 2014).

Nine manuals have been produced to cover the strategic investment sectors: Petrochemicals & Oleo-chemicals; Financial & Business Services; Creative Industries; Logistics; Tourism; Education; Health Services; Electrical & Electronics; and Food Processing & Agriculture. The manuals encourage private investors to introduce a more thorough and frequent monitoring system of environmental performance. Complementing this initiative, IRDA also established an Investment Committee in early 2014. The committee uses a checklist to ensure investment meets several criteria and to identify what might need improvement.

Source: OECD (forthcoming), *Green Growth in Iskandar Malaysia, Malaysia*; IRDA (2014), *Green Economy Guideline Manual*, http://iskandarmalaysia.com.my/downloads/IRDA-GEG-Manual-Creative.pdf.

Developing a skilled labour force for a green port city

A skilled labour force is a key factor in economic development, and in particular for attracting FDI in a global marketplace. In Viet Nam, various studies have indicated that an increase in skilled workers significantly increases FDI (Rowley and Warmer, 2013). Providing skills enhances people's capabilities, makes them productive and allows them to identify and take advantage of opportunities. Providing increased opportunities for skills development will help Hai Phong perform better. Despite an increase in its score, the results of a UNDP (2016) analysis in Viet Nam ranks Hai Phong among the weakest-performing provinces and cities in terms of human development, given its potential (its annual growth ranks 54th out of 63 provinces). To date, the central government and Hai Phong offer a number of programmes and vocational training courses aimed at developing human capital. Major initiatives include:

- **The National Target Programmes on Employment and Vocational Training** are two national programmes intended to enhance vocational training quality and effectiveness in order to create jobs and boost incomes of rural workers. They are devoted to labour and economic restructuring in the areas of agricultural, rural industrialisation and modernisation.
- Job creation loans through **the National Fund for Employment** provide small loans through its revolving credit fund. The fund advances approximately VND 200 billion (USD 9.46 million) of loans per annum to assist job creation at preferential rates.
- Overseas training programmes undertaken by the Government of Viet Nam to provide government scholarships for training to develop and strengthen the country's human resource capacity.
- Databases on the labour force play an important role in evaluating the economy and labour market demand, to provide enough human capital to respond to the changing needs of the market economy. Viet Nam's Ministry of Planning and Investment compiles such data.
- In Hai Phong, 11 vocational colleges, 10 vocational high schools and 24 job centres help improve workers' vocational skills. The Ministry of Labour, Invalids and Social Affairs is responsible for this strategy to develop Vietnamese human resources through the vocational school development project by 2020 (Decision #630 of 2012) and supports these institutions.

As Hai Phong's economy is specialised in port-related industries and a massive expansion and modernisation of its port is under way, the city's competitiveness can be increased by programmes providing **specialised job and skill training** associated with the **development and management of modern port-related facilities and equipment**, such as information and logistics systems including cargo-handling infrastructure. These initiatives could be strengthened by including jobs and skills related to environmental activities, such as consulting for clean equipment production, environmental impact assessment and research and development focused on technology to improve the port's environmental performance (see Chapter 3). Financial support to SMEs related to green businesses should also be encouraged, since Hai Phong's businesses are mainly SMEs with low innovation capacity.

Hai Phong's universities and institutes, such as the Institute of Marine Environment and Resources, the Institute of Seafood Research and the Institute of Marine Medicine, are important assets for the city. They should be used to promote capacity building, local support and engagement and innovative strategies, as well as to generate and accumulate local knowledge on safe environmental strategies to support green growth. They could also be a hub for collaboration and research among government, industry and academics to promote green growth.

Using economic zones and industrial clusters as "test-beds"

Hai Phong's economic zones (EZs) and industrial clusters (ICs) have attracted considerable foreign and domestic manufacturing investment. To date, seven EZs and five ICs have been developed, and accommodate about 150 companies. New EZs and ICs are also planned to the north and south-east of the traditional urban core in several satellite areas. EZ provides various tax incentives for investors that add to their appeal (Table 2.2). The **Hai Phong Economic Zone Authority** provides a "one-stop service" for investors in Hai Phong's industrial parks and EZs. The **Dinh Vu – Cat Hai Economic Zone** covers an area of 22 540 hectares (non-tariff zone 1 258 ha; tariff zone 12 532 ha: others 8 750 ha), extending to the north and south-east of the city. It is one of eight major coastal EZs in Viet Nam approved by the prime minister to focus on promoting investment with the state budget between 2016 and 2020.

While such EZs and ICs are expected to contribute to Hai Phong's economic development, they also provide untapped opportunities to promote the "greening" of Hai Phong. One possible option is to **designate a few EZs and ICs as "test-beds" and to pilot various instruments to encourage green growth.** For example, a comprehensive energy-efficiency package could experiment with promising co-generation, a waste-to-energy incinerator, a mechanism for sharing excess heat and electricity among enterprises and households near the IC, and incentives for SMEs and start-ups to introduce energy-efficient facilities. Policy design and implementation at the level of the EZs and ICs would allow a more comprehensive approach, rather than sectoral policies at the national or local level. Hai Phong could also introduce a mechanism such as a forum to facilitate knowledge exchange and diffusion on green growth among the EZs and ICs. Such exchange may be most efficient if foreign and domestic firms, as well as large and small firms, can interact with each other. Universities and Institutes could also play an important role in sharing knowledge.

Table 2.2. **Tax incentives in economic zones at Dinh Vu – Cat Hai Economic Zone, Hai Phong**

Tax	Incentives in EZ	Normal tax	Notes
Corporate income tax	Rate 10% for 15 years Exemption for 4 years Reduction of 50% for a period of 9 years	20%	From the first year, enterprises generate revenues from activities entitled to tax incentives. From the first year, enterprises generate earnings before taxes from investment projects.
Personal income tax	Reduction of 50%	100%	Individuals directly working in EZ
Import-export tax	Exemption tax according to the Decree No. 87/2010/NĐ-CP (for specific projects)	Decree No. 87/2010/NĐ-CP (for specific projects)	In non-tariff area
Value-added tax	Decree 209/2013/NĐ-CP (for specific projects)		In non-tariff area
Special consumption tax	Decree 108/2015/NĐ-CP (for specific projects)		In non-tariff area

Source: City of Hai Phong (2015a), "Answers to the OECD case study questionnaire", internal document, unpublished.

2.2. Ensuring successful urbanisation through land use and transport

As of 2011, 33% of Hai Phong's land was used for agriculture and 13.9% was forested (General Statistics Office of Viet Nam). Due to rapid urbanisation, such land is gradually being converted into urban land. The built-up area of Hai Phong increased by 18.1% (from 213.15 square kilometres to 251.65 square kilometres) between 2000 and 2010 – about 1.8% a year. It is observed that recent urban development has expanded from the south of Cam River to the south-east and west as well as north-west towards Ha Noi (Figure 2.1). Most of the inner-city areas are filled up by urban land use. It is estimated that 475 square kilometres of new land is required for urban development to accommodate 1.5 million new urban residents in the next 10 years (1.1 million new city residents and 0.4 million migrating from the rural districts of the city). How the city accommodates its increasing physical development needs will help determine the success of urban green growth.

Land use is always a critical factor for successful urban green growth in rapidly urbanising cities, because the urban physical form established during the process is likely to persist for decades. Getting a city right as it urbanises will be dramatically easier and cheaper than fixing things later. The risk of locking in the urban form should be addressed at the centre of urban green growth policy making. With the current outward expansion of the city, developing effective land use with an integrated transport system will be needed to avoid unplanned settlements and urban sprawl. An effective land-use policy would also enhance city resilience by steering development in Hai Phong away from risk-prone areas.

Figure 2.1. **Built-up area expansions in Hai Phong 2000-10**

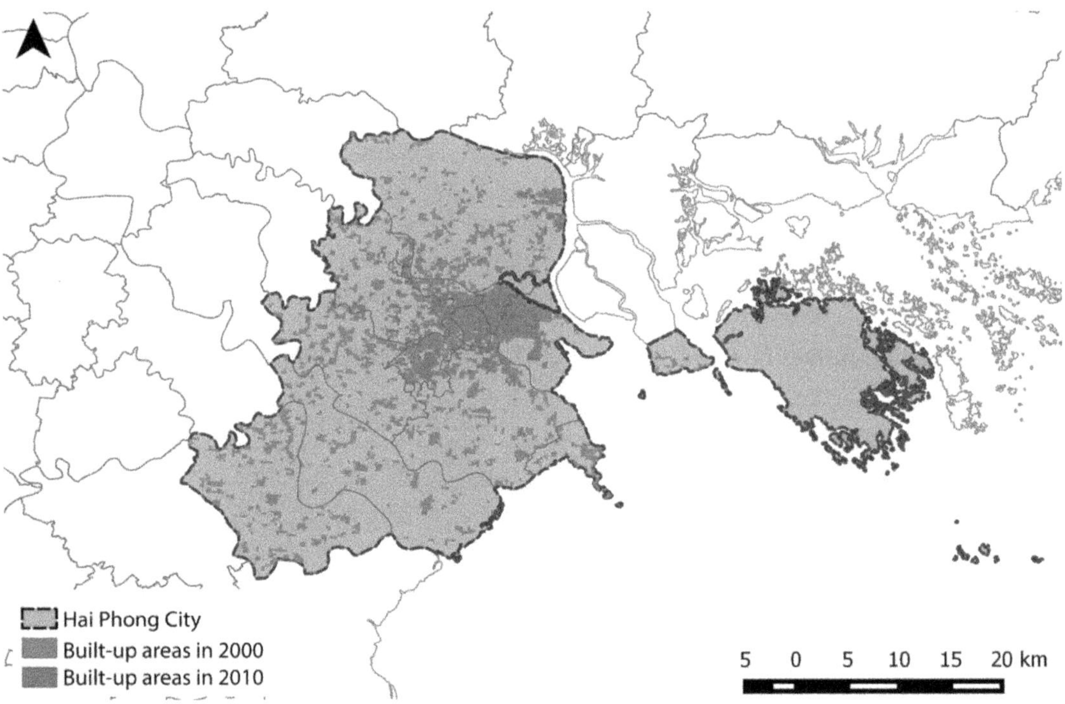

Note: In this analysis, the total land area of Hai Phong is calculated as 1 366.5 square kilometres.

Source: Calculations based on Mertes, C.M. et al. (2015), "Detecting change in urban areas at continental scales with MODIS data", *Remote Sensing of Environment*, Vol. 158, pp. 331-347, http://dx.doi.org/10.1016/j.rse.2014.09.023; Schneider, A. et al. (2014), "A new urban landscape in East-Southeast Asia, 2000-2010", *Environmental Research Letters*, Vol. 10(3), http://documents.worldbank.org/curated/en/844181467988870670/A-new-urban-landscape-in-East-Southeast-Asia-2000-2010 (accessed June 2016).

Guiding development with a clear land use vision and effective implementation mechanisms

A clear land use vision will help the private sector anticipate Hai Phong's long-term urban development and to invest in line with this vision. **The city master plan** is expected to facilitate this (Figure 2.2). Based on the Ministry of Construction's national urban master plan, the Hai Phong Department of Construction is responsible for preparing the city master plan. The target is 2050 and is required to be adjusted every five years. Based on this plan, the detailed Planning and Zoning Regulations (on a scale of 1/2 000 in 7 districts) are produced and approved by the People's Committee of Hai Phong.

If such a plan is to be effective, legal assurance is needed that the city master plan is aligned with all the long-term thematic and sectoral plans and strategies, including green growth. Major plans and strategies for Hai Phong include:

- **Hai Phong Socio-Economic Development Master Plan** is the main document for development of Hai Phong. Based on Ministry of Planning and Investment's National Socio-economic Development Master Plan, a local plan is prepared by the Department of Planning and Investment (DPI). It is a five-year plan, and sets targets for GDP growth, industrial production, capital expenditures, tax revenues, port cargo, job creation and poverty reduction, etc. Some targets for urban green growth are included, such as water and waste. However, there are a handful of long term plans which needs to be aligned with the city master plan. The current plan is for the period from 2011 to 2015. In addition, a 10-year strategy of socio-economic development for the city is in place for the period from 2011 to 2020.
- **Hai Phong Port Master Plan** is prepared at the national level, by the Ministry of Transport (MOT). In terms of spatial planning of the port, the city can propose its request to the MOT. Regarding the location and capacity, it is the responsibility of MOT.
- **The Green Growth Strategy Action Plan** was adopted in 2014, based on the National Green Growth Strategy and the National Action Plan on Green Growth for the period 2014-20. This action plan aims to green urban areas, green industry, green the environment and transform Hai Phong into a "green port city", by taking advantage of the historical port city and setting up a modern, competitive and eco-friendly green port system. In 2015, the city also produced the Green Growth Promotion Plan to identify tangible projects to implement the Action Plan.

Information provided for this case study suggests that the city master plan needs much clearer orientation for green growth as a reflection of these plans from a spatial perspective and communicate how urban development should proceed to realise all the other plans and strategies towards green growth. It is also indispensable for Hai Phong to formulate **a mobility master plan** to supplement the city master plan and guide the city to transit-oriented development. Essential elements for such a planning instrument have been identified according to international good practice (Box 2.3).

At the same time, **effective development control mechanisms** to guide the private sector to achieve the vision are necessary. Transparency and predictability in development control mechanisms will increase the credibility of the city master plan and help to promote private investment under the plan. Viet Nam has a unique land ownership system with Land-use Rights (LURs) (Box 2.4), and development control in urban areas is based on zoning. It has been argued that the process to obtain a land use certificate is not transparent and takes long time, which leads some private developers to escape from applying for a certificate. Effective enforcement is another urgent challenge (see Chapter 4).

Co-ordination among the city's departments should be enhanced to deal with conflicting land-use interests in Hai Phong. For example, under Decision No.1600/QĐ-UBND in 2013, the master plan for forest protection and development of the city by 2020 assigns 24 238.1 hectares as forest land by 2020. Although the city has been implementing forest development projects with this in mind, the forested area was only 17 990 hectares in 2013 (11.8% of the total area). Increasing competition for land among various business sectors, inadequate capital for the protection and regeneration of forests, especially mangrove reforestation, due to the limited city budget, and lack of co-ordination within the governments (among departments, districts and communes) are noted as major reasons (City of Hai Phong, 2015).

Figure 2.2. **Land-use plan for 2025 under the city master plan**

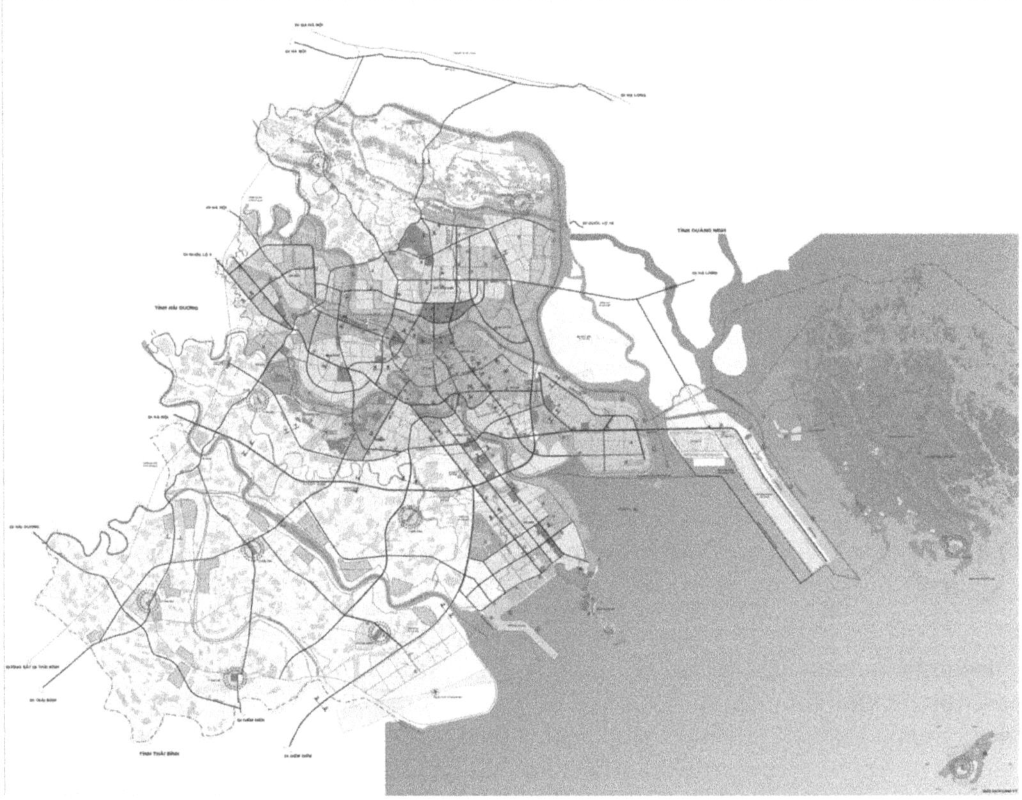

Source: Hai Phong (2015a), *Spatial Land use plan of Hai Phong 2025 to 2050*, http://haiphong.gov.vn/PortalFolders/ImageUploads/UBNDTP/932/Giao-thong-Quy-hoach-chung.jpg (accessed 20 July 2016).

Box 2.3. A common structure for developing successful integrated mobility plans

The International Union for Public Transport (UITP) develops *Action Points*, sets of recommendations in specific fields of public transport policy. As part of this work, the institution has analysed international experience in developing Integrated Mobility Plans and identified the following elements as essential for such planning tools:

- **Vision:** the long-term political vision of the city/region, including the future role of sustainable mobility.
- **Context:** setting the scene, detailing the transport geography of the area and studying the integration with land use and other policy areas, including plans for growth.
- **Challenges and strategic policies:** the challenges that need to be overcome are analysed through clear objectives and analysis of potential scenarios. Policies are developed to achieve these goals through the strategy.
- **Transport proposals:** sustainable multimodal transport proposals to achieve the strategy's objectives and overcome future challenges.
- **Expected outcomes:** analysis of the impact of transport proposals on reaching the objectives.
- **Implementation plan:** short-, medium- and long-term plans for delivery.
- **Costs and resourcing:** Identifying funding sources for the strategy.
- **Monitoring and reporting:** A framework of indicators and targets to measure the strategy's performance.

Source: OECD (2015), OECD Territorial Reviews: Valle de México, Mexico, OECD Publishing, Paris, http://dx.doi.org/10.1787/9789264245174-en. UITP (2014a), "Action points for the public transport sector. Connecting people and places. Integrated mobility plans for sustainable cities," Union Internationale des Transports Publics, Brussels, http://www.uitp.org/connecting-people-and-places-integrated-mobility-plans-sustainable-cities (accessed July 2016).

Box 2.4. Land-use rights in Viet Nam

Private ownership of land is not permitted in Viet Nam, and the people hold all ownership rights, with the state as the administrator. However, the laws of Viet Nam allow ownership of a right to use land. This right is called the **Land-Use Right ("LUR")**. There are three main regimes for investors to acquire LURs from the state:

- **Allocation:** the state can allocate LURs by administrative decision. Allocated LURs can be subject to a land-use fee or not, depending on the case.
- **Recognition:** the state can "recognize" LURs, in which case no fee is applicable.
- **Leasing:** the state can lease LURs on the basis of a contract. LURs leases are subject to a land-use rent.

With the Land Law 2003, the land market in Viet Nam was changed substantially from government control to a market-based system. The key features of the new Land Law 2003 are summarised as: the formal recognition of real estate markets, with a relaxation of government control of the land supply and cost; the devolution of responsibility from the central government to local governments in terms of land administration and registration; the use of market values for evaluating the compensation, taxes and land-use fees; public participation in planning; and the public notification of approved plans.

Source: Socialist Republic of Vietnam (2016), *Land regulations*, http://VietNamembassy-usa.org/basic-page/land-regulations (accessed 20 July 2016); Kato, H. and L.H. Nguyen (2009), *Land Policy and Property Price in Hanoi, Vietnam*, http://intl.civil.t.u-tokyo.ac.jp/docs/wpaper/Kato_Le_en.pdf.

Bringing jobs and homes closer and connected

Hai Phong occupies an area of 1 520 square kilometres and consists of a central city area, several satellite urban settlements in mountainous areas, two island districts and the Cat Ba archipelago. These urban areas are divided by six rivers flowing from the north-west to the south-east, where they reach the sea. The geographical distance between residential areas and job centres results in long commute times, traffic congestion and damage to public health due to air pollution. **Encouraging proximity and connectivity between jobs and homes** would help reduce these negative externalities and improve economic competitiveness, while improving the quality of life. Urban development needs to ensure that job centres have sufficient housing nearby, under the city master plan. Mixed-use development should also be promoted for this purpose. Making Hai Phong's industry greener can provide an opportunity to locate residential areas in closer proximity to industrial areas. Equally important are the connections between jobs and housing, especially in relation to public transport systems and good facilities for cycling and walking. Mixed-use development and "clean" connectivity should also target residential and commercial areas, in particular shopping centres, enhancing the quality of life by providing services in close proximity to a large number of residents. Recommended policy options include the following:

- **Setting density targets** for new development in the city master plan can effectively guide private development toward sustainability. Although excessive density may negatively impact the urban environment, sufficient density can help the city increase the efficiency of public services by lowering unit costs and using land more efficiently. In Hai Phong's land-use plan for 2025, density is utilised to calculate necessary land area for development (160 square metres per person) but this is neither a requirement for development permits nor a guiding principle for developers. Hai Phong could set density targets for new commercial and residential development and provide developers with incentives such as floor-area ratio bonuses. In addition, the city could introduce **minimum density requirements** as part of development controls, particularly in strategic locations such as the urban core and those located near public transport corridors.

- **Introducing transport accessibility assessment** as a requirement for large-scale development can guide major job and residential centres along well-serviced bus routes. The procedures for granting development permits will have to be linked to the assessment of the accessibility of the area's public transport network, based on both the city master plan and a mobility master plan. Development will thus be funnelled to areas covered by the existing transport infrastructure or where public transport infrastructure is planned under the mobility master plan.

- **Reserving land for necessary public services** in new development areas in accordance with the land use vision should be made a priority for Hai Phong. The cost to the city of making land available for public infrastructure, such as roads, public transport, parks and facilities for energy, water, wastewater and solid waste treatment, would be prohibitive as urbanisation advances. One option would be to create a supplementary plan under the current city master plan that identifies the land necessary for public services and could specify measures explaining how to acquire this land or make it available on a timeline. The city could also require private developers to provide land for public services or pay fees at the time of development. Such mechanisms are widely utilised in many OECD countries, such as Community Amenity Contributions (CACs) in Vancouver, Canada.

Redeveloping old and inner port areas will enhance the city's attractiveness and resilience

Since the Port of Hai Phong was built in 1874, Hai Phong has always been developing with the port. As port activities grew, the port expanded from Hoang Dieu (old port) towards the mouth of the Cam-Ca River, and the urban areas expanded along with the port areas. The port is now composed of several terminals scattered in and around the city. While the port brings the city economic prosperity, port activities including maritime and inland cargo traffic has negative environmental impacts including air pollution and noise, which can have severe consequences for the health of the population. Moreover, urban residents are suffered from congestion on urban road networks due to cargo traffic (Chapter 3).

Currently, the Port of Hai Phong is undertaking further expansion, including the construction of the new deep water seaport terminal of Lach Huyen, in order to respond to growing demand. This means many old terminals near the urban centre will be progressively phased out of the port system. Hai Phong should utilise this opportunity and **redevelop old and inner port areas and transform them into attractive urban and environment-friendly waterfront**, as experienced in many OECD port cities such as Hamburg (Germany), Yokohama (Japan) and Gothenburg (Sweden). High-quality urban redevelopment with mixed land use (residences, offices and commercial facilities) can bring people back to urban centres and attract tourists. Moreover, well-planned waterfront urban development can contribute to enhancing resilience of the city. Hai Phong can engage a wide range of stakeholders to develop a long-term vision of old and inner port areas and reflect it to the city master plan and other long-term plans and strategies.

Tackling a motorcycle-dependent urban environment

Vietnamese cities are one of the world's clearest examples of **motorcycle-dependent urban mobility**, and Hai Phong is no exception. Motorcycles have played a greater role in the motorisation of Asia than anywhere else in the world. Countries in the ASEAN region, including Viet Nam, have recorded strong growth in motorcycle ownership, which starts at very low income levels (OECD/ITF, 2013). Declining prices for motorcycles and the high social value associated with owning them in Asian countries partially explains the high levels of motorcycle ownership. As a result, many cities have become highly reliant on this type of vehicle for citizens conducting their everyday activities. This phenomenon is referred to as motorcycle dependency. The main characteristics of motorcycle-dependent cities are: i) high levels of motorcycle ownership; ii) lack of public transport alternatives; iii) an environment unsuited to bicycle transport and walking; and iv) a high share of motorcycles trips in total trips (Van et al., 2013). In 2009, motorcycles accounted for 93% of Hai Phong's vehicle fleet of 667 035, which indicates that there are over 620 343 motorcycles in the city (World Bank, 2011), which clearly demonstrates Hai Phong's dependency on motorcycles.

Motorcycle dependency is associated with high environmental and social costs. Users face a high risk of road accidents and fatalities. In 2010, over 70% of fatal crashes in Viet Nam involved motorcycles (Ivers, Nguyen and La, 2014). Motorcycles, and especially those with two-stroke engines, emit high quantities of pollutants per unit of activity, such as CO, NO_x and PMs (Dora and Hosking, 2012). Moreover, experience shows that **motorcycle use in a range of Asian cities may**

also accelerate urban sprawl. The fact that motorcycles can access areas with narrow and/or poor road infrastructure only supports sprawling forms of urban development. Without adequate spatial planning, lower density and scattered urban development may spread even faster than in car-oriented cities (Van et al., 2013). As a result, it is difficult for authorities to expand service provision to match the pace of urban expansion. In the case of public transport, the difficulty of providing bus services is complicated by development whose roads are unable to accommodate buses. The lack of access to public transport in many areas is further exacerbated by increasing motorcycle dependence.

To reverse motorcycle dependency and address traffic congestion, Hai Phong will need to make drastic and prompt **improvements to public transport**. For example, the city should ensure the implementation of its plan to put 220 eco-friendly buses into operation by 2020. Cities that have implemented ambitious public transport policies at early stages of development have been more successful in achieving more balanced mobility shares (Box 2.5). Even where motorcycle dependency is an intractable problem, **drastic improvements in public transport can lead to more balanced modal shares**. In the case of Taipei, the share of motorised trips by car and motorcycle was 76% in 1995. However, the city has significantly expanded its subway network since 1996 and introduced dedicated bus lanes that have helped increase the share of public transport from 24% in 1995 to 33% in 2010. An integrated payment system for different modes of public transport was an important part of the strategy (Tuan, 2012). Integrated e-ticketing was achieved with the introduction of a smart card (the Easy Card), which covered payment for all modes of transport and parking. In addition, significant discounts for public transport fares and transfers were granted to Easy Card users (Najman, 2008). In Taipei, use of private vehicles has been significantly reduced by improvements in public transport infrastructure and services. Demand management has also played a key role. For example, since 1999, motorcycles have no longer been permitted to park along sidewalks, and motorcycle parking pricing was introduced in certain areas in 2004.

It is also **crucial to prevent motorcycle dependency from becoming car dependency**, given citizens' strong preference for replacing motorcycles with cars.[1] The availability and quality of road infrastructure will play a decisive role in the extent to which the vehicle fleet shifts from motorcycles to cars as incomes grow. In a context of under-developed public transport, improvements in road infrastructure will transform motorcycle dependency into car dependency, but will not help manage overall motorisation rates. It is thus important for Hai Phong to plan upgrades and extend its road network in tandem with **effective demand management policies**. New road infrastructure should also take into account the need to prioritise road space for public transport. Examples from selected OECD and non-OECD cities exhibit complex policy options and their consequences (Box 2.5).

In the long term, within the balanced modal shares among different transport modes, motorcycles could play a more positive role in urban mobility if their use was limited to short trips and they were converted to **a feeder mode for public transport**. This strategy would have to be accompanied by strict regulation of motorcycle emissions and the introduction of road safety measures (Tuan, 2012). This strategy is also relevant for Hai Phong. In the short term, the role of motorcycles in meeting accessibility needs in Hai Phong needs to be acknowledged. Adequate regulation will be needed in order to avoid health and safety risks.

Box 2.5. **Tackling motorcycle and car dependency: Examples from OECD and non-OECD cities**

In Tokyo and Seoul, timely development of an extensive and well-integrated public transport network has made it possible to manage motorisation better. In Tokyo, the development of a dense rail network began in 1940, with the construction of the suburban rail system. The network was complemented with extensive development of the subway between 1950 and 1970 (Tuan, 2012). Tokyo's transit-oriented development is reflected in high public transport use and share of walking trips, despite its inhabitants' high incomes (Cheong and Loh, 2013). In Seoul, buses played a prominent role until the 1970s. Recent bus reform and the introduction of Bus Rapid Transit (BRT) have helped to reverse the diminishing role of buses for several decades. In addition to bus reform, the development of the suburban railway and the subway system have seen the public transport system attain a share of more than 50% of total motorised trips. In both Tokyo and Seoul, motorcycle ownership has never exceeded 50 vehicles per 1 000 population, and motorcycle shares have remained at 2-3%.

By contrast, Ha Noi and Jakarta show that a lack of public transport accelerates growth in motorcycle ownership and use. Once a city comes to rely on motorcycles, it becomes more challenging to attract users to public transport. In Jakarta, the BRT system introduced in 2004 has not managed to attract motorcycle users. Most riders had previously been using conventional buses and paratransit services. A similar phenomenon is noted in Bangkok. The BRT has probably meant better services and accessibility for these two groups, but it has not been able to reverse motorcycle dependency.

Introducing adequate demand management tools is also crucial in preventing motorcycle dependency from becoming car dependency. Both in cities like Bangkok and Tokyo, motorcycle ownership and use shifted significantly to car ownership and use as income levels rose. In Bangkok, policies to restrain the use of private cars were never implemented (although they were often proposed). In contrast, Tokyo has prevented car dependency with travel demand-management strategies. One of the most stringent is requiring buyers of cars to secure a night-time parking place before they are granted permission to register a car. This requirement is hard to meet, as parking is scarce and costly in the city (Cheong and Loh, 2013).

While car use typically displaces motorcycles in the long run, some cities have remained motorcycle oriented. In Taipei, a period of motorcycle growth was followed by a period of higher growth in the car fleet. However, this stabilised later on, while motorcycles experienced a second period of fast growth. A likely explanation is that motorcycles were more convenient, given the congestion resulting from high motorisation in a context of shortage of parking spaces and the absence of a hierarchical road network (Tuan, 2012). In Ho Chi Minh City, the motorcycle has remained the dominant mode of transport. Inadequate road infrastructure has been one factor in its unusual reliance on motorcycles, reducing the role cars can play in urban mobility (Van et al., 2013).

Source: Tuan, V.A. (2012), "Long-term strategies for motorcycle management in Asian Cities", *ITPS Transport Policy Studies' Review*, Vol. 14(4), pp. 72-80 (in Japanese); Van, Nguyen Thi Cam et al. (2013), "Urban accessibility in motorcycle dependent cities: Case Study in Ho Chi Minh City, Viet Nam", World Conference on Transport Research, 15-18 July 2013, Rio de Janeiro; Cheong, C.C. and L.O.H. Nadiah (2013), "Transport Policies and Patterns: A Comparison of Five Asian Cities", *JOURNEYS*, September.

2.3. Applying a comprehensive approach for water, solid waste and flood resilience

Keeping up with the increasing demand for drinking water

In Hai Phong, most citizens have access to treated water (98.91% of urban districts and 93.59% in rural districts as of 2014). In the urban districts and surrounding areas, clean water is supplied by Hai Phong Water Supply Joint Stock Company. There are two freshwater reserves (Trang Kenh, Nui Voi), and seven water supply plants currently producing 236 000 cubic metres daily. The water quality meets national technical standards for potable water (QCVN 01:2009/BYT) as well as the World Health Organization's standards (Finland Ministry of Foreign Affairs, 2013). A majority of suburban areas in the city are not yet connected to the centralised water system and rely instead on water supplied by mini water-supply factories without guarantee of its quality. Meanwhile, a long-term concern has been how to reduce dependence on underground water. The Hai Phong Water Supply JSC currently obtains about 34.6% of its water supply from other sources, including underground water and rainwater, to supply the rural and island district. Similarly, in Viet Nam, 30% of water supply is sourced from underground water and 70% from surface water (WHO, 2011). The National Centre for Water Resources Planning and Investigation reports that Viet Nam uses 5 million cubic metres of groundwater daily (VNS, 2016) and excessive ground water exploitation in some parts of the country has led to mineralisation content of over 1 000 mg/ml, high salt infiltration and land subsidence (Stefan, 2014).

With the increasing demand for water, an important challenge for Hai Phong is how to ensure that supply keeps up with demand. The city and water supply company need to continue to invest in water treatment plants and distribution systems to ensure stable water supply from local water resources. At present, Hai Phong is developing a water supply system expansion investment project financed by the Asian Development Bank. The total investment is VND 1.6 billion, and completion is expected by 2019, with additional daily capacity of 175 000 cubic metres. The project will ensure a constant supply of water to downtown areas and its vicinity from 2025. It is important to ensure the allocation of sufficient resources for such investment. In addition, Hai Phong should address **non-revenue water**, which stands at 13.7% of supply as of 2013. Thanks to Hai Phong's long-term efforts to reduce non-revenue water, the current figure is much better than the national estimated average of 35% (Canales-Trujillo et al., 2015) and the national target of 15%, as stipulated in 2009 (under the Decision No 1929/QD-TTg). Nonetheless, it remains high in comparison with other OECD cities.

Hai Phong should try to ensure adequate surface water supply and careful management of demand. Land subsidence due to ground water extraction should in particular be avoided, as it will increase the city's vulnerability to floods. Green infrastructure such as reservoirs and large-scale ponds, for instance, can be used to recharge local aquifers, capture surface water and channel it to treatment plants for supply to the population. At the household and building level, **rainfall harvesting systems** could be actively promoted and implemented in the city, and even made mandatory for new developments. Such water could be used for non-drinking purposes (e.g. toilets, bathing), helping to reduce demand for surface and ground water and easing pressure on local resources. Demand-side policies to curb consumption from local resources could also be explored. In Singapore, the implementation of the Four National Tap policy to tackle increasing water management issues was carried out alongside a Water Conservation Programme that has reached out to residents and the public and private sectors. Several initiatives have been launched in only a few years.[2] Economic instruments are also critical demand-side management tools: **block tariffs** could be used more aggressively to encourage large consumers to reduce consumption, while subsidies to the urban poor should be created to avoid negative externalities on poor families.

Investing in drainage and wastewater treatment systems

Hai Phong has no municipal wastewater treatment facilities at present. Poor surface water quality in Hai Phong poses serious environmental and public health hazards. When it floods, these risks can be amplified, mainly because of a lack of proper wastewater treatment combined with insufficient solid waste management. Studies have found that the Re River, Gia River and Da Do River, the major sources of surface water for the city, are polluted with levels of contaminants such as ammonia, iron and manganese as much as 3.8 to 9.9 times higher than the maximum limits (Viet Namnet, 2014). The contamination comes mainly from untreated domestic and industrial wastewater, excessive use of pesticides, fertilizer production, wastewater from landfills and cemeteries.

Urgent actions are required for industrial wastewater treatment. There are only six industrial zones which have local sewage treatment systems (Hai Phong People's Committee, 2016). Of 37 hospitals in the city, only 15 have a convergent wastewater treatment system, and the rest of the health facilities do not have a wastewater disposal system that meets standards (ibid.). Even where some local wastewater treatment stations have been built for industrial clusters, factories and hospitals, they are often not in operation, due to poor maintenance or simply in order to save the operation costs. A large proportion of wastewater is discharged directly into the environment (City of Hai Phong, 2015).

The constant pollution of surface water not only increases the cost of water treatment but increases the risk for the environment and human health, affecting drinking water and food production. The problem is exacerbated by seasonal floods, when the polluted water overflows into agricultural, forest and residential areas. The polluted rivers ultimately flow into the sea, with implications for coastal environments, including threats to the biodiversity of the 1 313 marine species found in the Cat Ba Archipelago.

Much needs to be done to improve the quality of surface water. A clear starting point is **periodical water-quality monitoring**. In Hai Phong, water-quality monitoring has been conducted since 2006 in channels, lakes, estuaries and rivers of the Re River, Gia River and Da Do River. However, the number of parameters covered and the frequency is not sufficient, given the limited city budget (City of Hai Phong, 2015). In addition, to strengthen pollution source control with industry, the city should invest in **installing real-time monitoring stations** in major water systems and make the data available to the public.

At the same time, urgent investment in the drainage system and wastewater treatment from houses and buildings in the city is needed. The sewer system of Hai Phong City, which is similar to other systems in urban areas across the country, is a tributary drainage system in which rainwater and wastewater flow through the same system. The major drainage system of Hai Phong's seven urban districts includes conditioning lakes (87.75 ha), pivot irrigation system (73.4 km), forced pump stations to drain rain water (2), sewage pumping stations (17), drain tide culverts (34) and over 550 km of pivot sewers.

The city is now constructing the first large-scale wastewater treatment plant in Vinh Niem Ward, Le Chan district, with a daily capacity of 36 000 cubic metres. More plants are planned in three districts (Hong Bang, Hai An, Ngo Quyen). Ensuring the construction and effective operation of these facilities will drastically improve surface water quality, but challenges remain in the areas of finance and citizen awareness. As Hai Phong relies on the private sector to provide many public services, investment in wastewater treatment plants also needs to be well co-ordinated with the private sector. Public-private partnership (PPP) experience in OECD cities could be useful in this regard. Wastewater treatment plant projects in The Hague, Netherlands and in Cairo, Egypt, represent successful use of PPP for the design, construction and operation of wastewater treatment facilities, with concessions of 30 and 20 years respectively. Proper management of user fees would be essential for financing the operation of the planned wastewater treatment facilities and to attract more investment for similar projects. The sewage treatment fee is relatively low (only 15% of the clean water price) (City of Hai Phong, 2015).

Improving management of hazardous and industrial waste

Solid waste management has been privatised in Hai Phong to operators selected by tender. Waste collection service covers all urban districts and approximately 90% of rural districts in the city. The Department of Construction (DOC) is in charge of domestic waste in urban areas, and the Department of Agriculture and Rural Development is in charge of domestic waste in rural areas. The volume of urban waste per person per day generated in Hai Phong has increased from 0.7 kg in early 2000's to 1.3 kg in 2015. The rates are expected to increase, owing to consumerism and the expanding economy (Nguyen Hoai Duc, 2014; City of Hai Phong, 2015). Increasing domestic and industrial solid waste generation is a threat to the environment and human health if not properly managed. Equally though, solid waste management presents green growth opportunities through recycling and energy generation. Hai Phong should improve solid waste management in two ways: minimise damage to the local environment and public health, while making the most of the potential for growth.

To address environment and public health concerns will require more effective management of hazardous and industrial waste. Major industrial waste includes: i) varnish products (paint, varnish, enamel glass), adhesives, sealants and printing inks; ii) waste oil, waste from liquid fuel, organic solvent waste, refrigerants and propellants; and iii) packing waste, sorbent, rags, filter materials, protective fabrics. In Hai Phong's industrial zones, such as Nomura, Dinh Vu, Trang Due, south of Kien Bridge and Do Sonthese, industrial waste is collected and treated (mostly by burning, while gas and oils are recycled) by five private companies. However, compliance remains a challenge. While there is a self-reporting system that requires registered firms to report the management of industrial waste four times a year, it is a major challenge for the city to regulate Hai Phong's more than 10 000 enterprises. In particular, SMEs have capacity problems for treating their waste properly. As with hazardous waste, rules are in force, but their implementation remains a challenge. For example, most city hospitals and some district hospitals have established environmental protection plans and annual environmental monitoring. Most hospitals have registered hazardous waste generators declared to the Department of Environment and Natural Resources. They have made contracts with Hai Phong Urban Environment Company on the collection, transport and disposal of medical hazardous waste. Two incinerator plants have been built for medical waste (supported by overseas development assistance, or ODA, from Austria and Japan). However, implementation is not always as expected, and hazardous medical waste is often mixed with domestic waste.

Options for Hai Phong include providing support for SMEs and smaller hospitals and clinics to improve their hazardous and industrial waste treatment and at the same time allocating more resources for reporting and inspection systems. For example, Hai Phong's "Master plan of medical waste treatment for the period of 2011-15, with orientations towards 2020" sets the targets outlined below, and its implementation needs to be carefully monitored:

- to invest in the construction of medical waste treatment systems with a view to substantially redressing environmental pollution at central and local medical establishments
- to enhance capabilities for environmental pollution observation and assessment at medical establishments, which will serve as a basis for elaborating plans for incremental improvements addressing environmental pollution at medical establishments
- to conduct scientific research into medical waste treatment
- to formulate and submit to competent authorities for approval four projects and schemes on: i) investments in waste treatment systems, ii) a master plan on hazardous medical solid waste treatment systems, iii) a project on capacity building for agencies in the health sector and iv) a scientific research scheme to intensify the application and transfer of advanced and environmentally sound medical waste treatment technologies.

Another important option would be to make **treatment fees** more transparent. In the current system, fees for hazardous and industrial waste are decided by negotiation between companies and service providers.[3] Solid waste management in the port needs also to be taken seriously. Various investigations at the port have led to the discovery of attempts to import hazardous materials that violate environmental regulations, such as lead-acid batteries, iron scrap, waste plastic and waste oil (Kojima and Michida, 2013). The Port Authorities should introduce a system (infrastructure and personnel) to detect hazardous materials such as these in the city and the country as a whole (see Chapter 3).

Promoting waste reduction and recycling

Sustainable waste management practices and the notion of the 3Rs (reduce, reuse and recycle) can help reduce costs and create new businesses. In particular, **a systematic approach to promote waste recycling will create opportunities for growth**. The domestic waste management of Hai Phong is reliant on landfills, with incineration accounting for a marginal amount of waste disposal. The city has five landfills in three districts (one in Tien Lang, two in Thuy Nguyen and two in Cat Hai) and 114 temporary landfills in the communes (DONRE, 2016). Few activities or policies promote recycling at present. While the seven urban districts of Hai Phong city generated 1 037.4 metric tonnes of waste per day in 2015, only 100 tonnes was composted and most recyclable waste was mixed with other waste and landfilled. In the EU, although 31% of waste is landfilled, a significantly higher share is recycled (28%) as well as composted (15%) (Eurostat, 2015). Singapore also recycles 60% of its waste. Recyclable materials are often picked up by scavengers from low-income households who separate and sell it on to recyclers (Pariatamby and Tanaka, 2014). While such actions are useful in the creation of jobs and enhance the livelihoods of low-income earners, separating such recyclable materials at the source is far more efficient. Moreover, composting in Hai Phong poses a difficult question. Even though several composting facilities at landfill sites have been experimented with (with the support of the Korea International Co-operation Agency, or KOICA, for example), many non-compostable materials are mixed in, which makes it difficult to produce good-quality compost (for use as fertiliser) despite the fact that a high proportion of waste generated is organic. The priority for Hai Phong is thus to separate recyclable goods at the source. At present, three urban districts are undertaking a waste separation pilot project (2014-18) namely, Lam Son, Tran Nguyen Han and An Duong. Such projects should be accelerated and scaled up to more parts of the city.

The landfill sites for Hai Phong could be more effectively managed to minimise negative environmental impacts and generate energy. The Fukuoka semi-aerobic technology (the Fukuoka Method), which permits rapid microbial decomposition and reduces methane gas concentration, had positive results when it was piloted at the Dinh Vu dumpsite. Applying low-cost technology, the piloting witnessed a carbon index drop from 30.4% to 8%, while the methane gas index dropped from 67% to 6% (Ngoc, 2014). The method reduces GHG emissions and is environmentally sound. Another option would be to capture methane gas generated from these landfills and utilise it as an energy source. The city can experiment with new technologies at the new landfill site planned at Gia Minh solid waste treatment complex in Thuy Nguyen district, which is 15 kilometres away from the centre and a part of the project on rain and wastewater drainage and solid waste management.

The city can provide technical assistance to SMEs so they can assess and optimise their production process to increase efficiency and reduce waste. Such efforts can directly improve the economic competitiveness of SMEs. Community-based action has played a key role in reducing the city's domestic waste by setting up targets and various pilot projects that empowered identifiable groups in waste collection and sorting for recycling. For example, in Cebu, small and medium-scale businesses have identified waste treatment and recycling as a business and located a market for their end products. Co-operatives were formed and received local government assistance to cover capital costs. Waste composting has created job opportunities and incomes for poor households under proper regulation through a local business approach. As a result, the total waste transported to landfill has been significantly reduced, by 16% (Premakumara, 2012). Another area of interest in reducing waste is to reduce reliance on plastic, since its rate of decomposition is slow. Paper bags and biodegradable plastics that are less harmful for the environment should be promoted across the city and supported by local legislation to increase their use. The city should also develop a waste management strategy to incrementally compost and recycle municipal waste.

Rural, by comparison with urban, areas in Hai Phong lag in the quality of their solid waste collection, transport and treatment. In rural areas, household waste is collected and transported to the landfills in the communes, the majority of which are temporary and the disposal methods unsanitary. Each year, the Department of Agriculture and Rural Development reports that about 4 500 tonnes of waste are generated, mainly by cattle and poultry feed, and about 200 000 cubic metres of emissions released to the air. The city's "**Tasks and solutions for collection and treatment of rural solid waste of Hai Phong City in the period of 2010-20**", a pilot initiative, has had considerable success in improving the management and treatment of rural solid waste (Box 2.6). This example could be disseminated to other urban and rural districts in Hai Phong and beyond.

Box 2.6. **Piloting solid waste management in rural areas in Hai Phong**

In rural areas of Hai Phong, solid waste treatment mainly involves sending to landfill and does not yet meet environmental technology standards. To limit pollution and environmental degradation, **"tasks and solutions for collection and treatment of rural solid waste of Hai Phong City in the period of 2010-20"**, was initiated by the Department of Agriculture and Rural Development in 2010, to improve the management and treatment of rural solid waste. The following targets were set:

- by 2015: 70% of solid waste, 70% of hazardous industrial solid waste, 80% of hazardous solid waste in medical services, and 90% of industrial non-hazardous solid waste will be collected and handled to the appropriate environmental standards
- by 2020: 90% of solid waste will be collected and treated to environmental standards, and 100% of industrial solid waste and hazardous solid waste in health care services will be handled in accordance with regulations.

To achieve the targets, a number of policy instruments were introduced. The department supported the districts to prepare and submit applications for the approval of the planning locations for concentrated solid waste treatment stations and garbage transfer stations. Based on the approved master plans, the districts were able to actively invest in the construction of concentrated treatment stations. Then, the department provided equipment and moved all the communal solid waste to the newly constructed concentrated treatment stations, and closed the communal landfills. Each district was encouraged to conduct research and propose solutions for efficient collection and treatment. Waste collection fees were also introduced, which should in principle cover the cost of waste collection in the communes. The industrial production companies, trade villages, clinics were required to make contracts of waste collection and treatment in accordance with general regulations.

The achievement in the period 2010-15 was remarkable. By June 2015, 97% (139/143) of all communes had established solid waste collection teams, and some communes had signed contracts with collectors and solid waste transport companies. The workforce from the communes who were involved in waste collection and transport numbered approximately 1 200 people. In 2012-13, 19 training classes were held, with over 2 660 attendees (those directly in charge of collecting, transporting and treating solid waste in the communes). Fifty transfer stations were newly built or renovated, and 86 temporarily solid waste landfills of communes created. A total of 1 252 handcarts and 300 waste bins were provided for the communes and investment was made for 6 incinerators for communes in the districts of An Lao, Kien Thuy, Thuy Nguyen, Tien Lang and Vinh Bao, equipped with BD-Alpha technology with a capacity of 500kg per hour.

In 2014, the ratio of rural waste collected and treated reached 81% (217 000 out of 267 830 tonnes), which exceeded 11% of the target. The percentage of waste in the rattan village, carpentry village, ornamental floriculture village collected and processed was 85%, which exceeded by 15% the goals laid out in the City Resolution.

Source: City of Hai Phong (2015), "Answers to the OECD case study questionnaire", internal document, unpublished.

Enhancing resilience to floods

Hai Phong has the highest density of rivers in Viet Nam's Northern Plain, with six major rivers traversing the city from the north to the sea. The delta area accounts for approximately 85% of the city's area, which spreads out in such communes as Vinh Bao, Tien Lang, An Lao, Kien Thuy, An Duong, and the South of Thuy Nguyen. The delta area stands between 0.7 metres-1.7 metres above the sea level (City of Hai Phong, 2015). The city is vulnerable to flooding on a periodic basis, and tsunamis at any time, as well as being threatened over the longer term by sea-level rise, localised flooding due to more intense rainstorms (precipitation) and ocean storm surges. These risks are due in large part to Hai Phong's location on the coast of Viet Nam, which is subject to intense tropical storms (typhoons) and weather patterns (monsoons). They are exacerbated by population growth in risk-prone areas with inadequate climate adaptation capacity or protective measures. The "Continuity Regional Business Plan" project, collaboration between Hai Phong and the Japan International Co-operation Agency (JICA) in 2012-14, identified the following as the main types of natural disasters affecting Hai Phong:

- storms and tropical depressions formed on the Pacific Ocean, the South China Sea or the Gulf of Tonkin, and hitting land over the territories of Viet Nam
- heavy rain concentrated in the region
- flooding caused by upstream floods, heavy rain, storm surges and tropical depressions in the estuaries and coastal areas.

These threats have the power to severely damage Hai Phong's critical infrastructure and built environment, not to mention the port. The result will have repercussions on regional trade, financial investments and the provision of basic services. Those most likely to be directly affected by flooding are the poor and vulnerable sub-populations of women, the elderly, the young and disabled, living downstream in densely populated, flood-prone areas. However, the indirect impacts of flooding will be experienced much more broadly; both by local people affected in economic terms (through lost employment or livelihoods) and lost trade with markets in northern Viet Nam and southern China.

According to city records, 43 storms and typhoons (tropical depressions) directly affected it between 1990 and 2015. The statistical data in recent years show a rise in the number of storms that affected coastal defence. Typhoon No. 8 (*Sơn Tinh*) in 2012 caused extensive damage in 10 northern cities, including Hai Phong, where it was considered the most destructive in the last 10 years. The citywide property losses were estimated at VND 1 trillion. Damage has also been inflicted by tornados, droughts and landslides (City of Hai Phong, 2015).

Hai Phong's policies for disaster-risk reduction (DRR) and disaster-risk management (DRM) are based on **the 2008 Action Plan** for implementing the National Strategy for Prevention and Mitigation of Natural Disaster with the vision 2020, based on the National Strategy (Box 2.7). Such a plan, often referred to as a **Local Resilience Action Plan (LRAP),** would certainly be useful, although a LRAP serves no purpose if it is not implemented through specific concrete programmes, budgets and actions. It is essential that implementation takes place before disasters strike, since prevention is the key to success.

Effective preparation for, and quick recovery from, future floods in Hai Phong will require **both "hard" and "soft" infrastructure**. In terms of hard infrastructure, targeted investment in the drainage system should be considered in flood-prone neighbourhoods. The city's drainage capacity (culverts, canals and conditioning lakes) is insufficient and unlikely to withstand disasters involving more than 100 millimetres of rainwater (City of Hai Phong, 2015). This creates serious environmental and health risks given the inadequate wastewater and solid waste management system. Inadequate maintenance, and the possibility of accidental releases of unmonitored toxic chemicals and waste from industrial and port areas, require upgrading of the city's storm drainage systems. Other "hard" infrastructure elements include eco-based adaptation measures such as retention ponds, creation of public open spaces or zoning restrictions in flood-prone corridors, early warning systems for floods, business continuity plans (BCPs), and disaster insurance and funds. Green infrastructure in particular includes adaptive solutions that allow "making space for water", a concept now widely adopted in flood-prone countries like the Netherlands. In Portland, Oregon, the city's 2005 Watershed Management Plan (PWMP) uses plants and soil to slow, filter and infiltrate runoff close to its source, in a way that strengthens and mimics natural functions/processes (OECD, 2012). Malmö (Sweden) has set up a system to drain rainwater from rooftops and other impermeable surfaces and channel it through canals, ditches, ponds and wetlands into a sub-surface conventional sewer system to avoid overburdening the traditional drainage network (Neumann et al., 2015).

The key for investment in resilience is to recognise **complementarities and synergies with other policy objectives**, especially in association with the natural environment, and to promote measures that serve multi-dimensional outcomes. For example, a green space used as a retention pond in case of floods can be used as an urban park for citizens and to protect local ecosystems and biodiversity. Such space can also function as a buffer zone between industrial parks and residential neighbourhoods. In addition, it would help to recharge local aquifers by draining run-off into the ground and to purify water, ensuring water supply and sanitation services (OECD, 2015b). A comprehensive, integrated **floodwater management plan** could inform land use, wastewater treatment and other policies related to floodwater management and provide more co-ordinated and effective solutions. Such a plan should consider both the threat of flooding from inland, upstream areas that flow into Hai Phong, as well as coastal flooding caused by sea-level rise, erosion typhoons or storm surges. Wider co-ordination with surrounding provinces would be critical for such a plan to be effective.

Preserving and restoring natural ecosystems should be considered an important resilience measure. Cat Ba Archipelago Biosphere Reserve plays an important role in carbon sequestration and helps to reduce GHG from building up in the atmosphere. It is home to about 2 320 wildlife and plant species, some of which are endangered. The site also attracts a significant number of tourists yearly. It is imperative that the natural reserve be regularly monitored, to avoid any adverse impact from the harbour and ship industries. The depletion of mangrove sites for shrimp aquaculture increases the vulnerability of the city, eliminating the natural defence system that reduces the impact of storms. The successes at Bang La show that with collaboration from local communities, the mangrove sites can be well managed and conserved. Hai Phong could also fund programmes to support climate-smart agriculture to capture carbon in soils and vegetation, improve yields and resilience to droughts and floods, and protect water quality and biodiversity. Inter-sectoral synergies will be needed to implement these programmes.

In terms of governance, Hai Phong appears to have developed a co-ordinating mechanism for DRR/DRM. **The Steering Committee for Natural Disaster Prevention and City Rescue** is a cross-departmental organisation that takes responsibility for advisory, planning management, general disaster-risk management, search and rescue, and oil spill reaction. When a disaster occurs, based on the functions and tasks of the branches and units, the Committee is responsible for co-ordination, creating favourable conditions for the effective collaboration of all agencies in the response activities. Members of the committee are heads of Hai Phong departments, agencies and government units. Every year, they are asked to prepare local plans for natural disaster prevention and submit them to the committee so it can prepare a master plan for the whole city.

Hai Phong also recognises that citizen engagement and the role of the private sector are key factors in successful DRR/DRM. Based on a national project for raising public awareness and community-based risk management of natural disasters, Hai Phong developed an action plan for 2013-15 (Box 2.7). It is also implementing the Disaster Response Project, a collaborative project between the city and Seattle, sponsored by the USAID. Its focus is on improving enterprise resilience and improving public-private collaboration in planning for disaster response (Chapter 3).

Box 2.7. **Plans and strategies for resilience in Viet Nam**

In 2007, the central government produced the **National Strategy for Prevention and Mitigation of Natural Disasters with the Vision 2020** (Decision No. 172/2007/QD-TTg regulated by the Prime Minister on 16 November 2007). This aims to mobilise resources to put into effect up to 2020 disaster prevention, response and mitigation to minimise loss of human life, property and damage to natural, cultural and environmental assets, in the interest of the country's sustainable development and national security. Based on the strategy, Hai Phong produced the Action Plan on implementing the National Strategy for Prevention and Mitigation of Natural Disaster with the vision 2020 (No.5295/KH-UBND on 12/9/2008, regulated by City People's Committee).

The central government also launched a project for **raising public awareness and risk management of natural disasters based on community resource** (Decision No. 1002/QD-TTg on 13/7/2009, approved by the prime minister).The programme targeted 6 000 communes facing risk of natural disasters and offered training to key staff from all provinces in disaster reduction and raising awareness of potential risks. Based on this programme, Hai Phong drew up an action plan for the period 2013-15 to implement the national initiative (Plan No.5935/KH-UBND on 14/8/2013, regulated by the city's People's Committee). The local action plan aimed to make the communities of Hai Phong and its people more resilient in the case of natural disaster and to the effects of climate change.

Source: City of Hai Phong (2015), "Answers to the OECD case study questionnaire", internal document, unpublished.

2.4. Greening the energy and building sectors and promoting green products

The energy sector in Viet Nam presents substantial green growth opportunities, as well as challenges. National electricity power produced by power plants and imported from foreign countries in the year 2015 was 161 026 billion kWh, an increase of 10.47% over 2014. The national maximum electricity power (Pmax) in 2015 was 25 290 MW (City of Hai Phong, 2015). In 2010, the national GHG inventory indicated that energy was the largest contributor, emitting 141.1 million tonnes or 53.05% of total GHG emissions (Ministry of Natural Resources and Environment, 2014). This was followed by agriculture, accounting for 33.20% (88.3 million tonnes). Industrial processes and waste respectively accounted for 7.97% (21.2 million tonnes) and 5.78% (15.4 million tonnes). Since 1994, emissions from the energy sector had risen by a factor of more than five, particularly between 2000 and 2010. This highlights the pressing need for more sustainable renewable energy sources (Figure 2.3).

Hai Phong's electricity consumption pattern is typical for an industrial city. Sixty percent of the electricity is consumed by the industrial sector, 36% by households, and 3% by the commercial sector. Energy in Hai Phong is supplied by hydro and coal-based thermal power, imported from China through the national grids. Two co-generation plants are located in Hai Phong, but there are no power plants. The energy supply is apparently still keeping up with increasing energy demand. The current capacity in 2015 in Hai Phong was about 3 670 million kWh, whereas the actual demand is around 2 130 million kWh (the surplus can be sold commercially). Demand, however, is expected to increase by 13.7% (2016-20).

In 2010, the total GHG emissions in Hai Phong were 17 207 103 tons of carbon dioxide, and emissions from the energy sector accounted for 76.9% (Figure 2.4). By 2020 under a business as usual situation, energy share of GHG emissions would increase to 79.2%. Similarly, the share of industrial processes and water, sewage and waste emissions would increase, whilst the share of agriculture, and transport emissions would decrease. Manufacturing industrial energy alone accounts for the highest share (65%) of GHG emission, and is projected to increase by 284% in 2020 (City of Hai Phong, 2015). Given its rapidly increasing energy consumption and GHG emissions, Viet Nam urgently needs to take action at all levels of government. While the national government is responsible for structural energy policies to transform the energy mix by creating a market environment in favour of cleaner energy sources, large cities like Hai Phong can contribute to national goals by piloting renewable energy options, promoting greener products and energy efficient buildings, in addition to improving energy efficiency in industry. Many local energy policies can create co-benefits, such as reduced air pollution and improved traffic congestion, which correlate effectively with local actions.

Figure 2.3. **GHG emissions by sector in Viet Nam (1994-2030)**

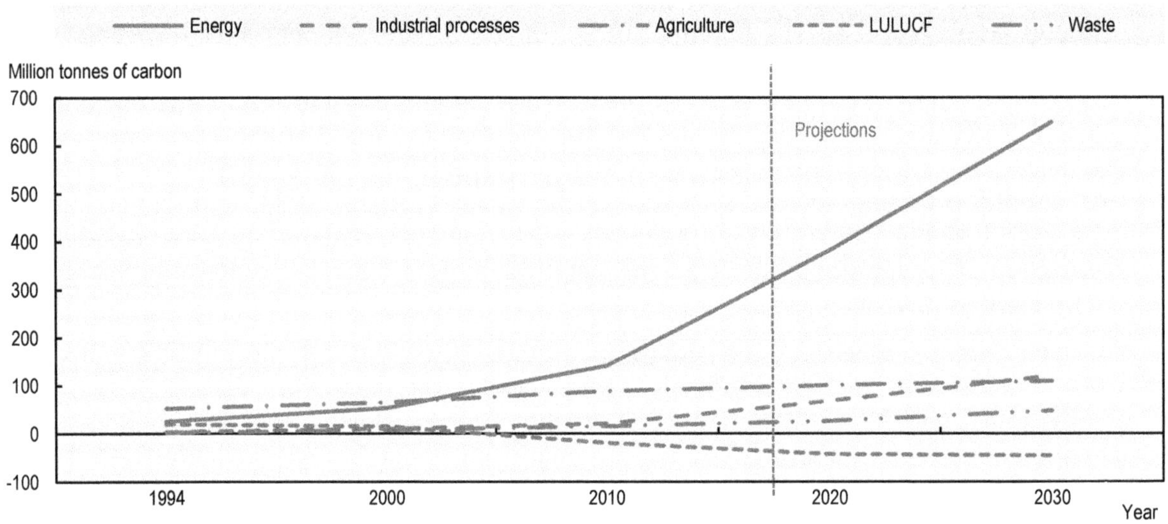

Source: Ministry of Natural Resources and Environment (2014), *Biennial Updated Report of Viet Nam to the United Nations Framework Convention on Climate Change*, http://unfccc.int/resource/docs/natc/vnmbur1.pdf.

Figure 2.4. **GHG emissions in Hai Phong by sector**

2010 and-2020

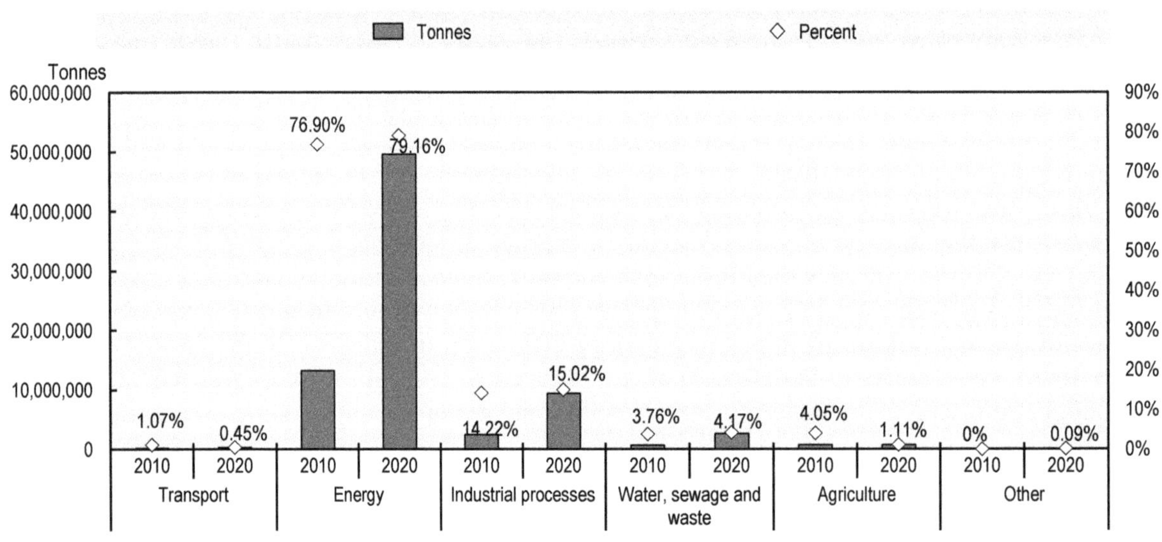

Source: City of Hai Phong (2015b), Green Growth Promotion Plan of the City of Hai Phong, Hai Phong, Viet Nam, http://www.asiangreencamp.net/pdf/green_en.pdf.

Exploring renewable energy potential

Hai Phong has been exploring renewable energy options. To increase the proportion of its energy sourced from sustainable sources, use of biomass electricity is expected to rise from 0.6% in 2020 to 1.1% in 2030 (City of Hai Phong, 2015). In addition, the city is promoting solar hot water heaters and providing financial support to citizens purchasing new solar-energy appliances. The city is also implementing pilot projects using solar batteries for public lighting and traffic lights (City of Hai Phong, 2015). Based on assessments of the pilot projects, these initiatives could be scaled up.

The city could also explore other untapped renewable energy potential, including **waste-to-energy technology**. Given that Hai Phong sends most domestic solid waste to landfill, modern incinerators with emissions control technology could significantly reduce emissions (Pyper, 2011). However, the development of a mechanism of waste separation would be a prerequisite for effective operation.

Wind and solar energy remain untapped sustainable energy sources in Hai Phong. Earlier attempts to install wind turbines in 2004 demonstrated the renewable sector's great potential.[4] According to the city, although there is substantial variation between seasons, there are good average wind speeds. The annual average wind speed is 2.4 m/s, while higher average wind speeds of 3.4 m/s have been recorded in July. During the south-east monsoon, which occurs over the summer, wind speeds average 2.50 m/s, with maximum speeds of 20-30 m/s. The results of a survey by the Mining, Geology and New Energy Centre recommend further exploitation of solar and wind energy. The average annual amount of sunshine experienced by the city is over 1 600 hours, with radiation upward of 1 150 kwh/m^2/year. In terms of wind energy, the intensity differs by locality, but Bach Long Vi, for instance, has wind energy of 3200 kWh/m^2/year at a height of 10 metres, and up to 7 600 kwh/m^2/year at a height of 60 metres (ECCH, 2014b). This potential should be further exploited.

Fossil fuel subsidies must be progressively reduced to incentivise the use of cleaner energy sources. Fuel subsidies in Viet Nam are substantial. Between 2007 and 2012, they cost the government between USD 1.2 billion and USD 4.49 billion annually (UNDP, 2014). Fuel subsidies are regressive and deprive governments of the financial resources for socio-economic investment. The central government should urgently pursue **its roadmap to phase out subsidies for fossil fuels**, as laid down in the 2012 National Green Growth Strategy (NGGS).

Promoting green products

Another potential area for action is **energy labelling**, as a way of encouraging citizens to use more energy-efficient products. The practice varies widely in different countries. In Thailand, a green-label programme introduced by the national government for consumers and industrial products has been awarded to over 500 products (OECD, 2015a). Singapore has a green labelling programme that has certified 3 000 products (Singapore Environment Council, 2016), recognised across the region as meeting stringent environmental standards. In Viet Nam, a green label programme has existed since 2009, but its full potential has not been reached. The NGGS includes energy labelling as follows: i) labelling energy-saving equipment, issuing national standards for the quality of equipment; and ii) developing and issuing standards for economic sectors and green/eco-labelling of products. The central and local government should collaborate on an effective framework for green labelling. This could be piloted in Hai Phong and expanded across the country. If Viet Nam's green labelling programme were extended, the products under its banner could aim to achieve international recognition for high environmental and sustainability standards.

This programme could be linked with **green** public **procurement programmes** (at both central and local levels of government). Compulsory environmental specifications could be imposed on such goods and services, or extra points granted for green solutions, making it more likely that they are selected. In Thailand, green labelling schemes, aside from helping consumers identify and buy eco-friendly services and products, also encourage competition between manufacturers to produce greener products and services. Increasing public awareness of these services and products is also essential.

Promoting energy-efficient buildings

The building sector is also an important area for curbing energy consumption. New buildings have the greatest potential for high energy efficiency, but promoting the renovation and retrofitting of existing buildings is also worthwhile, since many older office and large residential buildings with lower energy efficiency ratings are likely to be renovated. The Department of Construction is responsible for enforcing building codes. The government of Viet Nam, through a partnership with the U.S. Agency for International Development (USAID), has updated its Green Building Code, which will promote high-performance and energy-efficient buildings in Viet Nam's major urban areas.

National and local governments should work together to promote energy-efficient buildings. The **green building labelling/certificate** is an initiative that Hai Phong could consider. Thailand's Green Building Institute (TGBI), the main agency delivering green building standards, also promotes the Thailand Rating Energy and Environment System (TREES), which provides advice and standards to building managers through a certification programme covering standards in energy consumption, materials and resources, and indoor environmental quality (OECD, 2015a). Such mechanisms could be used more effectively if they were combined with regulations and procurement frameworks as a part of building codes. Higher standards could promote the use of energy-saving appliances and energy-efficient materials in buildings, which also creates potential for technological innovation. A reporting mechanism for building energy efficiency could also be considered. Incentives such as floor area ratio and open space ratio bonuses will also be crucial for Hai Phong to ensure the application of the QCVN 09:2013/BXD, a National Technical Regulation on energy-efficient buildings. This standard is now applied for new construction with a ground floor area of 2 500 square metres or more and holds great prospects for energy saving.

In addition, the passive cooling of new buildings shows great potential. In hot climates, low-cost solutions such as reflective roofs and walls, exterior shades, and low-emissivity window coatings and films can significantly reduce energy consumption for cooling (IEA, 2013). National and local government institutions, with technical assistance from university and research institutions, could support research on passive cooling designs, and provide vocational training for a new generation of installation technicians and technical assistance for contractors and construction companies.

Hai Phong can also collaborate with the national government to enhance the skills of construction workers. This is crucial to ensure the effectiveness of all the policy instruments. Such a training programme could also can increase job opportunities and promote green growth. Hai Phong is participating in the "Viet Nam Clean Energy Program: Energy Efficiency Promotion in the Building Sector". This is a four-year pilot programme under the management of Ministry of Construction, supported by US Agency for International Development (USAID). Other participating cities are Ha Noi, Da Nang, Ho Chi Minh and Can Tho. The objective is to reduce electricity consumption and thus GHG emissions from the building sector. The programme also aims to support the implementation of a Green Growth Action Plan in Viet Nam's building industry. The programme provides comprehensive support for participating cities to collect data and develop technical standards for energy-efficient construction (NTR 09: 2013/MOC) and to promote green building technologies through market incentives. The city can collaborate with the construction sector in the city and scale up such actions. In Chicago, the city has managed to reduce emissions and waste while conserving energy and resources through cross-sector activities involving residential, municipal and commercial buildings. As of 2014, the process had helped save over USD 10 million and created numerous jobs. The programme included retrofitting old homes and green roofs (City of Chicago, 2016).

Smart technologies have the potential to be used in helping Hai Phong shape its green growth transition. Aiming for environmental resource efficiency and a competitive economy in a rapidly evolving world requires technological innovation as such, Hai Phong on an incremental basis could introduce smart technologies in its transport management system, carbon monitoring systems, renewable energy and e-governance, similar to Bandung's implementation of their smart city initiatives (Box 2.8). Wide stakeholder participation in this drive for innovation from business, civil society, academia and research would be essential.

Box 2.8. **Bandung Smart City**

Bandung, Indonesia, has emerged as a major smart city in the country. Boosted by its economic prosperity in the manufacturing and creative industries, the city's population is rapidly growing resulting in a number of urban challenges such as traffic congestion, waste management and pollution. To contribute to a better management of the city, a Command Centre was set up in 2015 with modern equipment to provide visibility across the city through the CCTV networks capturing real-time activity in the streets. GPS tracking systems are also employed to provide information in monitoring traffic and public assets.

Similarly, the city, pursuant to an e-government initiative, is providing digital government services to the citizens. This has been made possible by the internet access for all through the over 10 000 free Wi-Fi access points. Over 320 apps have been built, whilst social media monitoring helps government stay in touch and gain insights into the problems faced by over 2 million of its online users. The interactive citizen reporting system enables citizens to send in tweets, SMS or report online to draw the municipality's attention to a particular issue. The relevant department/agency is then assigned to the issue. Duration for problem resolution serves as an important performance metric. The smart city approach has helped contribute to improve the perceived quality of life of citizens.

In all, Bandung's Smart City application can be seen in the areas of infrastructure, smart and government, citizen engagement and innovative clusters. OECD's assessment of the city identified a greater potential and use of Bandung's smart technology in the area of urban resilience and disaster management especially in tackling the city's perennial floods.

Source: OECD (2016, forthcoming), *Green Growth in Bandung, Indonesia*; OECD (2015c), *Knowledge Sharing Workshop on Smart City and Green growth, Indonesia*, 6-7 May 2015.

Main policy recommendations

- Install **real-time monitoring stations (air and water)** for major pollutants at major industrial zones (IZ), industrial clusters (IC) and the port, and make the information accessible to public. Set targets for major pollutants and prepare action plans to meet the targets at every IZ and IC.
- Provide **technical guidance and financial support** to manufacturing industries, in particular to SMEs, to enable them to increase energy efficiency and gradually eliminate outdated and polluted production technologies.
- Provide **specialised job and skill training** for the development and management of modern port-related facilities and equipment. Provide financial support to green businesses.
- **Designate a few economic zones (EZs) and industrial clusters (ICs) as "test-beds"** and pilot various instruments to encourage green growth.
- Align **Hai Phong city master plan** with all the other thematic and sectoral plans and strategies including green growth, while ensuring effective implementation mechanism. Generate an **integrated mobility plan** to supplement the city master plan.
- Introduce **density targets** and **transport accessibility assessment** as a requirement for large-scale development. Create a supplementary plan under the city master plan that **identifies the land necessary for public services** and specifies how to acquire the land or make it available on a timeline.
- Engage a wide range of stakeholders to develop **a long-term vision of old and inner port areas**, transforming them into attractive and resilient urban waterfront.
- Invest in **public transport**, while introducing demand management tools to prevent motorcycle dependency from becoming car dependency.
- Invest in **drainage and wastewater treatment systems**. Provide support for SMEs and smaller hospitals and clinics to improve **their hazardous and industrial waste treatment** and at the same time increase resources for reporting and inspection systems.
- Expand the city's **waste separation pilot project** to more parts of the city. Provide technical assistance to SMEs so that they can optimise their production process to increase efficiency and reduce waste.
- Produce a comprehensive, integrated **floodwater management plan**.
- Explore untapped **renewable energy** potential, including waste-to-energy, wind and solar energy. Present **a roadmap to phase out subsidies for fossil fuels**.
- Introduce an **energy labelling** programme for electric appliances, linked to public procurement. Introduce a green reporting mechanism for building energy efficiency and establish **a green labelling system** for buildings.

Notes

1. In general, those with middle and high incomes prefer to own cars, which confer higher status, and the introduction of Tata Nano cars, for example, in many markets has allowed those with even lower income to shift from motorcycles to cars (Tuan, 2012). Furthermore, individuals may decide to change from having a motorcycle to a car for safety reasons (Van et al., 2013).
2. For more information see: www.pub.gov.sg/conserve/WACProgramme/Pages/default.aspx (accessed 10 July 2016).
3. For domestic waste, a fixed price list is set by the Ministry of Finance, with a subsidy available.
4. In 2004, Hai Phong collaborated with the private sector to build the first advanced wind power electricity station in Viet Nam, with 800 kW capacity, on Bach Long Vi Island. The successful construction of a wind power electricity station was intended not only to ensure a stable power supply for Bach Long Vi Island District but to help solve the long-term electricity issue for the offshore islands in Viet Nam. However, after a year of operation, the station's generator was severely damaged in a storm, and has not been operative since then (City of Hai Phong, 2015).

Bibliography

Asia-Pacific Economic Co-operation (APEC) (2009), *Viet Nam: Water and Waste Treatment*, http://egs.apec.org/more-articles/163-Viet Nam-water-and-wastewater-treatment (accessed 10 July 2016).

ADB (2015), "Greater Mekong Subregion Economic Cooperation Program: Overview", www.adb.org/sites/default/files/publication/29387/gms-ecp-overview-2015.pdf.

Canales-Trujillo, N., V.X.H. Vu and S. Whitley (2015), "Mapping current incentives and investment in Viet Nam's water and sanitation sector: Informing private climate finance", Overseas Development Institute, London, https://www.odi.org/sites/odi.org.uk/files/odi-assets/publications-opinion-files/9642.pdf.

Choi, C.C. and L.O.H. Nadiah (2013), "Transport Policies and Patterns: A Comparison of Five Asian Cities", *JOURNEYS*, September, https://www.lta.gov.sg/ltaacademy/doc/13Sep069-Loh_ComparisonOfFiveAsianCities.pdf.

City of Chicago (2015), *Retrofit Chicago*, www.cityofchicago.org/city/en/progs/env/retrofit_chicago.html (accessed 20 July 2016).

City of Hai Phong (n.d.), *City of Hai Phong Website*, http://haiphong.gov.vn/Portal/Default.aspx?Organization=ubndtp_EN (accessed July 2016).

City of Hai Phong (2015), "Answers to the OECD case study questionnaire", internal document, unpublished.

DONRE (2016), "Report of the environmental protection and environmental monitoring results in 2015; tasks and solutions in 2016", People's Committee of Hai Phong, No. 253/BC-UBND, Department of Natural Resources and Environment, 24 November 2015.

Dora, C. and J. Hosking (2012), "Urban Transport and Health: A Review", *JOURNEYS*, Issue 9, November, https://www.lta.gov.sg/ltaacademy/doc/J12%20Nov-p07Carlos_Urban%20Transport%20and%20Health-A%20Review.pdf.

ECCH (2014a), "Activities of Hai Phong Energy Conservation and Cleaner production Centre associated with green growth and sustainable development", Energy Conservation and Cleaner Production Centre, http://ecchaiphong.gov.vn/defaulte.aspx?pageid=newsdetailsE&catID=6&id=131 (accessed 12 June 2016).

ECCH (2014b), "Hai Phong promotes renewable energy", http://ecchanoi.gov.vn/default.aspx?page=home&lang=1&portlet=46&cat=92&content=1782 (accessed 12 June 2016).

Eurostat (2015), "Environment in the EU". http://ec.europa.eu/eurostat/documents/2995521/6757479/8-26032015-AP-EN.pdf/a2982b86-9d56-401c-8443-ec5b08e543cc (accessed 20 July 2016).

EX Corporation (1996), "Kitakyushu - a case study", *Metropolitan Environmental Improvement Program (MEIP)*, World Bank, Washington, DC, http://documents.worldbank.org/curated/en/1996/04/441745/japans-experience-urban-environmental-management-vol-2-4-kitakyushu-case-study (accessed 20 July 2016).

Finland Ministry of Foreign Affairs (2013), "Water Supply and Waste Water Treatment", Embassy of Finland, Ha Noi: Development Co-operation, last updated 23 July 2013, www.finland.org.vn/public/default.aspx?nodeid=39697&contentlan=2&culture=en-US (accessed 25 June 2016).

General Statistics Office (n.d.), *General Statistics Office of Viet Nam Website*, www.gso.gov.vn/default_en.aspx?tabid=466&idmid=3&ItemID=12830 (accessed June 2016).

Hai Phong (2015a), *Spatial Land use plan of Hai Phong 2025 to 2050*, http://haiphong.gov.vn/PortalFolders/ImageUploads/UBNDTP/932/Giao-thong-Quy-hoach-chung.jpg (accessed 20 July 2016).

Hai Phong (2015b), "The city's index of industrial growth in 11 months rose 16.55%", http://haiphong.gov.vn/Portal/Detail.aspx?Organization=ubndtp_en&MenuID=3292&ContentID=79782 (accessed 8 April 2015).

IEA (2013), Transition to Sustainable Buildings: Strategies and Opportunities to 2050, IEA, Paris, http://dx.doi.org/10.1787/9789264202955-en.

IRDA (2014), *Green Economy Guideline Manual*, http://iskandarmalaysia.com.my/downloads/IRDA-GEG-Manual-Creative.pdf.

Ivers, R. Q., H.T. Nguyen, and Q.N. La (2014), "Status of road safety and injury burden: Vietnam", *Journal of Orthopaedic Trauma*, Vol. 28, S50-S51.

Kato, H. and L.H. Nguyen (2009), *Land Policy and Property Price in Hanoi, Vietnam*, http://intl.civil.t.u-tokyo.ac.jp/docs/wpaper/Kato_Le_en.pdf.

Kojima, M. and E. Michida (eds.) (2013), *International Trade in Recyclable and Hazardous Waste in Asia*, Edward Elgar Publishing, Northampton.

Mertes, C.M. et al. (2015), "Detecting change in urban areas at continental scales with MODIS data", *Remote Sensing of Environment*, Vol. 158, pp. 331-347, http://dx.doi.org/10.1016/j.rse.2014.09.023.

Ministry of Natural Resources and Environment (2014), *Biennial Updated Report of Viet Nam to the United Nations Framework Convention on Climate Change*, http://unfccc.int/resource/docs/natc/vnmbur1.pdf.

Najman G. (2008), *Taipei: an example of Integrated e-Ticketing to favour intermodality*, CODATU XIII Ho Chi Minh City, www.codatu.org/wp-content/uploads/Taipei-an-example-of-Integrated-e-ticketing-to-favour-intermodality-G%C3%A9rard-NAJMAN.pdf.

Neumann, J. E., K.A. Emanuel, S. Ravela, L.C. Ludwig, and C. Verly (2015), "Risks of coastal storm surge and the effect of sea level rise in the Red River Delta, Vietnam", *Sustainability*, Vol. 7(6), pp. 6553-6572, , http://www.mdpi.com/2071-1050/7/6/6553 (accessed June 2016).

Ngoc, T. (2014), "Viet Nam Business Forum – VCCI", Vccinews.Com, http://vccinews.com/news_detail.asp?news_id=30704 (accessed June 2016).

Nguyen Hoai Duc (2014), *Waste Management in Vietnam – Potential for power generation*, presentation at GIZ Workshop, Ho Chi Minh City, 21 May 2014.

OECD (forthcoming), *Green Growth in Iskandar Malaysia, Malaysia*, OECD Publishing, Paris.

OECD (forthcoming), *Green Growth in Bandung, Indonesia,* OECD Publishing, Paris.

OECD (2015a), *Green Growth in Bangkok, Thailand*, OECD Publishing, Paris, http://dx.doi.org/10.1787/9789264237087-en.

OECD (2015b), *Water and Cities: Ensuring Sustainable Futures*, OECD Publishing, Paris, http://dx.doi.org/10.1787/9789264230149-en.

OECD (2015c), *Knowledge Sharing Workshop on Smart City and Green growth, Indonesia*, 6-7 May 2015.

OECD (2013), *Green Growth in Kitakyushu, Japan*, OECD Publishing, Paris, http://dx.doi.org/10.1787/9789264195134-en.

OECD (2012), *Compact City Policies: A Comparative Assessme*nt, OECD Green Growth Studies, OECD Publishing, Paris, http://dx.doi.org/10.1787/9789264167865-en.

Pariatamby, A. and M. Tanaka (2014), "Municipal solid waste management in Asia and the Pacific Islands", *Environmental Science*, Springer, Singapore.

People's Committee of Hai Phong (2015), "Environmental protection and environment monitoring results 2015", Report 253/BC-UBND, 24 November 2015.

Premakumara, D.G.J. (2012), "Establishment of the community-based solid waste management system in Metro Cebu, the Philippines", https://pub.iges.or.jp/pub/establishment-community-based-solid-waste (accessed 10 June 2016)

Pyper, J. (2011), "Does burning garbage to produce electricity make sense?", *Scientific American*, Environment & Energy Publishing, 26 August 2011, www.scientificamerican.com/article/does-burning-garbage-to-produce-energy-make-sense (accessed 13 June 2016).

Rowley, C. and M. Warner (2013), *Whither South East Asian Management?: The First Decade of the New Millennium*, Routledge, New York.

Schneider, A., C.M. Mertes, A.J. Tatem, B. Tan, D. Sulla-Menashe, S.J. Graves, N.N. Patel, J.A. Horton, A.E. Gaughan, J.T. Rollo, I.H. Schelly, F. Stevens and A. Dastur (2014), "A new urban landscape in East-Southeast Asia, 2000-2010", *Environmental Research Letters*, Vol. 10(3), http://documents.worldbank.org/curated/en/844181467988870670/A-new-urban-landscape-in-East-Southeast-Asia-2000-2010 (accessed 30 June 2016).

SCMP (2014), "Huge number of rubbish bins in Hong Kong is contributing to waste crisis, says activist", www.scmp.com/news/hong-kong/article/1670243/huge-number-rubbish-bins-hong-kong-contributing-waste-crisis-says (accessed 7 July 2016).

Singapore Environment Council (2016), http://sgls.sec.org.sg/index.php (accessed 18 March 2016).

Socialist Republic of Vietnam (2016), *Land regulations,* http://VietNamembassy-usa.org/basic-page/land-regulations (accessed 20 July 2016).

Stefan, C. (2014), "Groundwater vulnerability in Viet Nam and innovative solutions for sustainable exploitation", *Journal of Vietnamese Environment*, Vol. 6(1), pp. 13-21.

Themelis, J. N. (2006), "Solid waste management in Viet Nam: An industrial ecology study by Thao Nguyen School of International and Public Affairs", University of Columbia, New York.

Tuan, V.A. (2012), "Long-term strategies for motorcycle management in Asian Cities", *ITPS Transport Policy Studies' Review*, Vol. 14(4), pp. 72-80 (in Japanese).

UNDP (2016), "Growth that works for all: Viet Nam Human Development Report 2015 on Inclusive Growth", www.vn.undp.org/content/vietnam/en/home/presscenter/pressreleases/2016/02/05/viet-nam-human-development-report-2015-comprehensive-institutional-and-policy-reform-needed-for-growth-that-works-for-all.html (accessed 20 July 2016).

UNDP (2014), "Green growth and fossil fuel fiscal policies", www.vn.undp.org/content/vietnam/en/home/library/environment_climate/green_growth_and_fossil_fuel_fiscal_policies_in_viet_nam.html (accessed 20 July 2016).

Van, Nguyen Thi Cam et al. (2013), "Urban accessibility in motorcycle dependent cities: Case Study in Ho Chi Minh City, Viet Nam", World Conference on Transport Research, 15-18 July 2013, Rio de Janeiro.

Viet Nam National Green Growth Strategy (2012), Decision No. 1393/QD-TTg of the Prime Minister, https://www.giz.de/en/downloads/VietNam-GreenGrowth-Strategy.pdf.

Viet Namnet (2014), "Pollution contaminates Hai Phong waterways", http://english.Viet Namnet.vn/fms/environment/113054/pollution-contaminates-hai-phong-waterways.html (accessed 17 July 2016).

VNS (2016), "Experts warn coastal groundwater overused", http://vietnamnews.vn/environment/281417/experts-warn-coastal-groundwater-overused.html (accessed 6 April 2016).

WHO (2011), *Water and Sanitation Sector Assessment Report Viet Nam 2011*, World Health Organization, www.wpro.who.int/Viet Nam/topics/water_sanitation/watsan_sector_report_Viet Nam_2011.pdf.

World Bank (2011), *Vietnam - Haiphong Urban Transport Development Project,* World Bank, Washington, DC, http://documents.worldbank.org/curated/en/2011/02/13887684/vietnam-haiphong-urban-transport-development-project (accessed July 2016).

Chapter 3

Green Growth in the port of Hai Phong

Chapter 3 examines the green growth performance of the Port of Hai Phong. The port is a major driver of Hai Phong's growth, thanks to the trade and economic activities it generates not only locally but regionally. In addition, the port is currently experiencing rapid growth, which means increasing maritime and industrial activities, demand for infrastructure, potential environmental challenges and exposing an increasing number of assets to risk. This chapter is structured into the following four sections:

1) An introduction to the growth context of port activity in Southeast Asia and Hai Phong

2) Boosting the international competitiveness of the Port of Hai Phong

3) Greening the activities of the Port of Hai Phong

4) Making the Port of Hai Phong resilient to natural disaster risk.

This chapter benefited from discussions at the fourth OECD Knowledge-Sharing Workshop on Urban Green Growth in Dynamic Asia, held in Hai Phong on 25-27 June 2015 and supported by the OECD Knowledge Sharing Alliance.

Main points

- **The Port of Hai Phong is the largest seaport of northern Viet Nam**, accounting for 25% of total container throughput in the country in 2014 and second only to the port of Ho Chi Minh City. It has recently experienced rapid growth: cargo throughput increased from 7.7 million tonnes to 41.4 million tonnes between 2002 and 2014. The port directly benefits from Hai Phong's location, its main competitive advantage: the city is part of an important triangular economic zone in northern Viet Nam and lies at the heart of two economic corridors connecting to two dynamic southern Chinese provinces. The context of growth and increasing demand creates specific port challenges in terms of competitiveness, environmental impact due for instance to congestion, and resilience to typhoons and floods. These specific challenges and associated recommendations to encourage green growth in the port must be integrated with Hai Phong's Green Growth Promotion Plan and Action Plan, if the current vision of a green port city is to be fully realised.
- The Port of Hai Phong needs to become **more competitive** to increase its market share, respond to growing demand and to keep pace with other emerging ports in the region. While the construction of the new deep water seaport terminal of Lach Huyen should increase the maritime connectivity of the port, it is not performing well by other measures of competitiveness. Hinterland connections are poor because of the congestion created by an over-reliance on trucks for cargo transport (around 90% of the cargo transport modal split), which threatens to reduce trade volumes and values. The fragmentation of the port between more than 10 terminals and the lack of labour capacity to meet the demand of a fast-growing and modernising seaport are a drag on the efficiency of port operations.
- Several studies have noted a variety of **environmental impacts** related to the activities of the Port of Hai Phong. Maritime and inland cargo traffic result in the emission of pollutants such as CO, SO_2, NO_x, particulate matter and also dust. Levels of PM_{10} in Hoang Dieu, Chua Ve and Tan Vu terminals are on average 142 $\mu g/m^3$, 136 $\mu g/m^3$ and 141 $\mu g/m^3$, respectively. In addition, an estimated 3 000 to 5 000 tonnes of waste oil spilled during regular bunkering operations are generated in the port annually, only 20% to 30% of which is collected. Finally, several studies have reported significant loss of biodiversity due to low seawater quality and sedimentation, and handling of contaminated masses during dredging activities present further environmental risks. This is of great concern, as Hai Phong is located in an estuary with valuable and sensitive ecosystems of mangrove forests, coral reefs, seagrass and aquaculture.
- **Risk of typhoons and floods** is a potentially critical obstacle to green growth in Hai Phong, as explained in Chapter 1. The port's facilities are also at risk. Past natural disasters, such as the 2012 Son Tinh typhoon, destroyed or damaged ships, ship cranes, marine management stations and containers. Given the importance of the Port of Hai Phong in the regional economy and its concentration of infrastructure, its resilience to typhoons and floods is a key concern. The local government and the port authority need to improve the city and the port's adaptive capacity, considering that more intense storms may occur as a result of climate change.

3.1. Introduction: Opportunities and challenges of a fast-growing seaport

Rapid growth in regional port activity

The increasing centrality of Asia's port activity is symptomatic of its economic growth: in 1972, approximately 40% of all world port activity took place in Europe, 20% in North America and 20% in Asia. These shares had dramatically changed by 2009, when more than half of world port activity took place in Asia, around a fifth in Europe and a tenth in North America (OECD, 2014). At a lower scale, the Southeast Asian region illustrates this growth in port activity: according to UNCTAD Maritime Transport, the total fleet in Southeast Asia increased from 58 280 in 2002 to 150 337 in 2014. The number of container ships has increased from 5 630 to 22 069 (Maritime Insights, 2014). Container throughput has therefore increased dramatically in the region, and in 2013, the ASEAN-10 countries[1] accounted for 13.5% of the world's total container throughput. While Viet Nam accounted for only 1.2% of the world's total container throughput, container throughput in its ports increased around tenfold between 2000 and 2014 (Figure 3.1).

Figure 3.1. **Container throughput in selected ASEAN countries**

(2000-14)

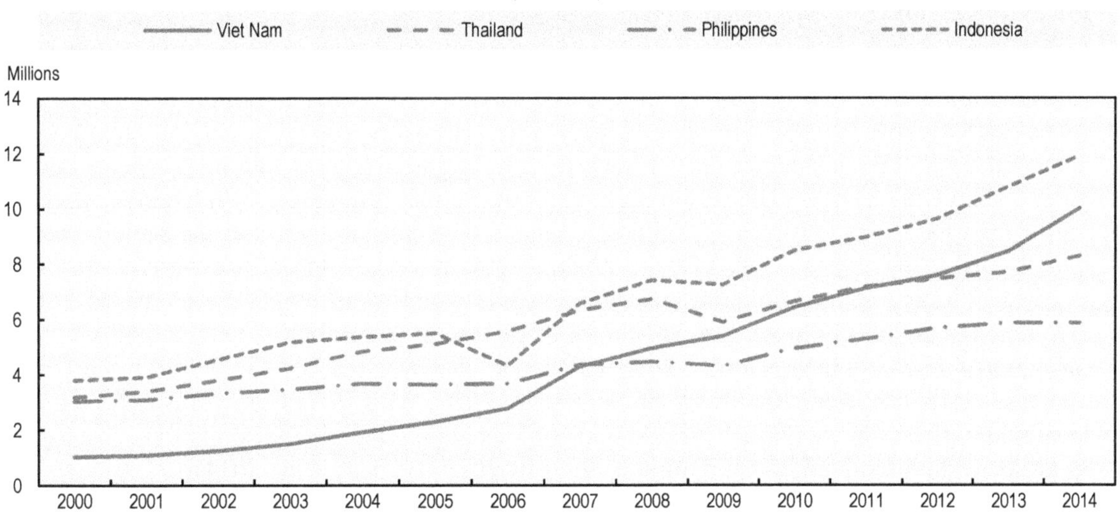

Source: World Bank (2014), "Container port traffic", *World Development Indicators*, http://data.worldbank.org/indicator/IS.SHP.GOOD.TU/countries (accessed 20 July 2016); and Viet Nam Seaports Association (n.d.), www.vpa.org.vn/english/information/info_static2014.html (accessed 23 June 2016).

The Port of Hai Phong: Dramatic growth driven by an advantageous location

The Port of Hai Phong was built in 1874, more than 140 years ago, during the French occupation. It is now composed of several terminals scattered in and around the city, and mainly along the Cam-Ca River. Together, these terminals form the largest and busiest seaport of northern Viet Nam, accounting for 25% of total container throughput in the country in 2014 (2.55 million 20 foot equivalent unit, or TEU, out of a total of 10 million TEU), and the second busiest seaport in the country after the port of Ho Chi Minh City (Dynaliners, 2015). Hai Phong is ranked 9th among Southeast Asian container ports and 69th in the world in terms of volume (Figure 3.2). Port activities are thus an important driver of local and regional economic development, with 37 companies exploiting the diverse terminals – 12 dealing with petrol, LPG and oil products, 10 with containers and 15 with general cargo. In addition, 117 companies are operating in the field of maritime services and 17 companies specialise in shipbuilding and repair.[2]

Figure 3.2. **Volumes of container port traffic in the main Southeast Asian ports**

(2014)

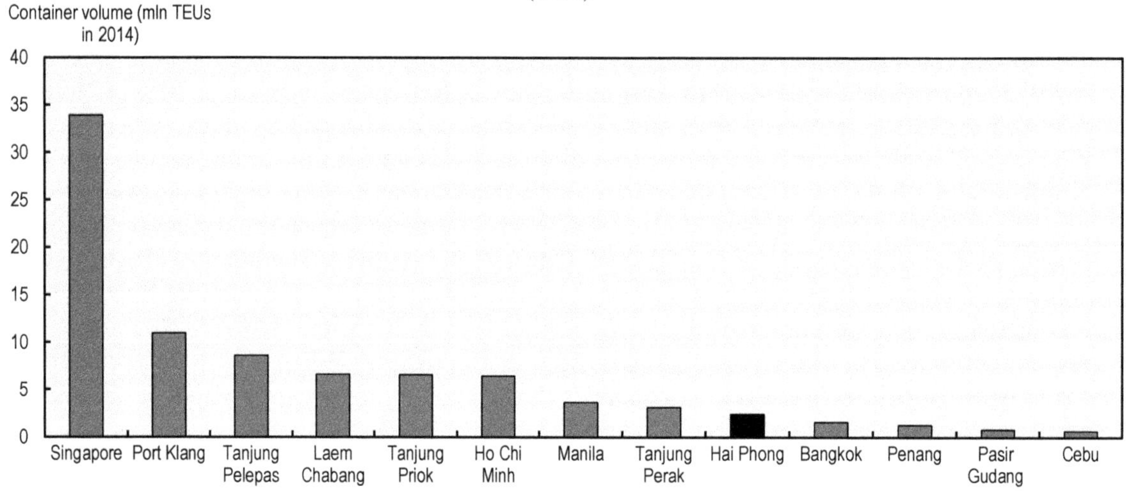

Note: Volumes for Cebu refer to 2013

Source: International Transport Forum/OECD collection data.

The Port of Hai Phong has experienced significant growth since the turn of the century: container throughput has increased from 219 200 TEUs in 2000 to 2.55 million TEUs in 2014 – an increase of 19.16% per year (Figure 3.3). Between 2005 and 2014, Hai Phong experienced a 316% container growth rate, the third-highest growth rate in Southeast Asia after Cebu City (390%) and Iloilo City (387%) in the Philippines.[3] In parallel, cargo throughput has increased from 7.7 million tonnes to 41.4 million tonnes in the same period, mainly driven by imports and domestic trade – although the volume of exports is almost equivalent to imports and domestic trade (Figure 3.4). Total wharf length has increased from 3 kilometres in 2000 to more than 11 km.[4] In 2014, export turnover was USD 3.5 billion, an increase of 18.3% compared with 2013, and import turnover reached USD 3.56 billion, an increase of 16.35% from 2013 (City of Hai Phong, 2015). This rate of growth is expected to continue. Demand in the northern region ports of Viet Nam – mainly the ports of Hai Phong – is projected to reach 5 million to 6 million TEUs by 2020 (Blancas et al., 2014).

Figure 3.3. **Container throughput in the Port of Hai Phong**

TEUs (2000-14)

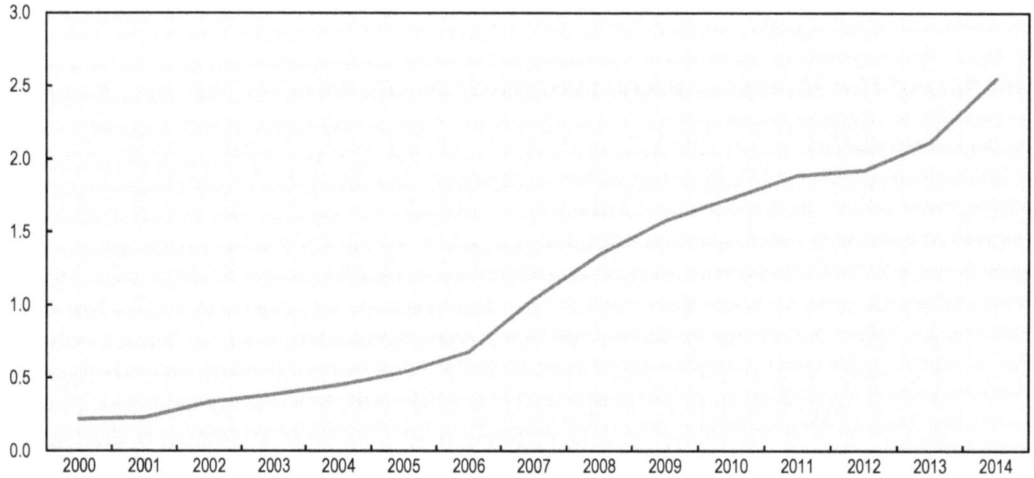

Source: Viet Nam Seaports Association (n.d.), www.vpa.org.vn/english/information/info_static2014.html (accessed 23 June 2016).

Figure 3.4. **Cargo throughput in the Port of Hai Phong**

Million tonnes (2000-14)

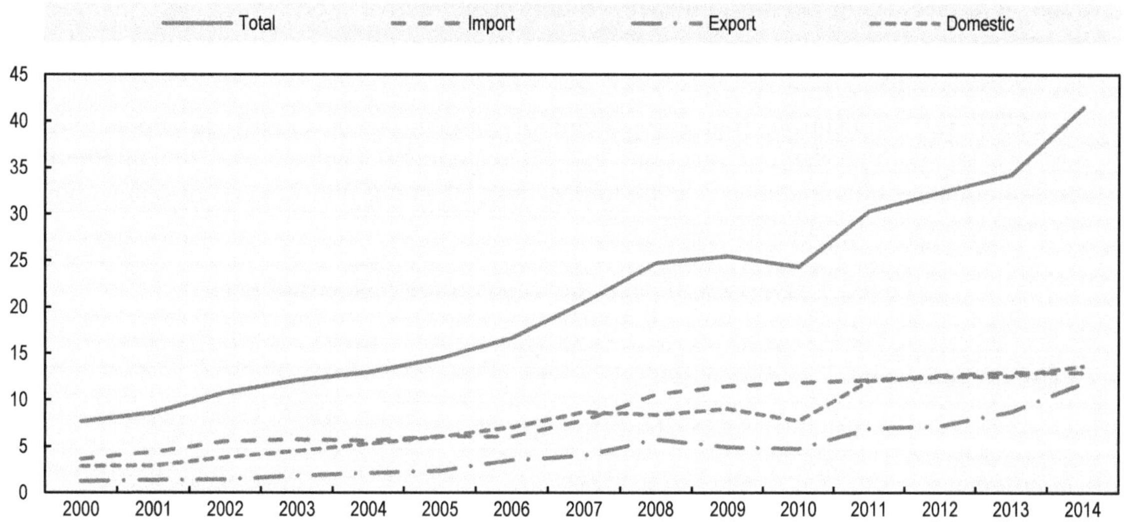

Source: Viet Nam Seaports Association (n.d.), www.vpa.org.vn/english/information/info_static2014.html (accessed 23 June 2016).

Hai Phong's geographical location is the port's main competitive advantage. The city is part of a triangular economic zone in northern Viet Nam (including Ha Noi, the country's capital, located only 100 kilometres away, and Quang Ninh), and lies at the heart of two economic corridors: one connecting Hai Phong to Kunming province (southern People's Republic of China [China]) through Ha Noi, and the coastal corridor from Ho Chi Minh City to Nanjing province (south-eastern China). Of goods arriving at the Port of Hai Phong, 90% are destined for transit through Ha Noi, including 15% that go to Southern China (Blancas et al., 2014).[5] Some of the main export products passing through the Port of Hai Phong are garment, footwear, aquaculture, plastic products, electric appliances and cables. Hai Phong is also a major recipient of foreign direct investment (FDI) in Viet Nam: it has benefited from over 450 FDI projects with total registered capital of USD 11 billion, many of them directly or indirectly benefiting the port.[6]

The growth of the port has presented several challenges, including the need to accommodate growing demand and modernise infrastructure, and the need to mitigate the potential environmental impacts associated with increasingly intense port activity. The region's exposure to disaster risk, in particular to typhoons and floods, is high. The port's resilience must be enhanced to avoid short-term and long-term damage to the local economy and infrastructure. These challenges should be addressed by both the local government and the national government. The national government, in particular, must acknowledge the importance of the Port of Hai Phong for northern Viet Nam including Hanoi and support efforts to promote green growth in the Port.

3.2 Boosting the international competitiveness of the Port of Hai Phong

The dramatic growth experienced by the Port of Hai Phong has brought an array of challenges, including deficits in infrastructure, the development of port operations, quality of hinterland connections and labour skills. If not addressed, these will undermine the port's competitiveness. Given the pace of development, the port is at a crossroads and needs to seize the current opportunity to transform itself into a major international port. The stakes are all the higher given that the Asia-Pacific region is now the world's main trade interface, with well-established, highly competitive ports such as Shanghai, Singapore and Hong Kong, and numerous ports gaining traction in the emerging markets of Southeast Asia.

OECD (2014) identifies four components of competitiveness: maritime connectivity, port operations, hinterland connections and local goodwill (Table 3.1). For Hai Phong, trans-shipment, skills mapping and matching, port planning and hinterland connections are critical levers of competitiveness.

Table 3.1. **Overview of policy instruments for competitive ports**

Determinant	Instrument	Examples
1) Maritime connectivity	Trans-shipment	Singapore
	Nautical access	Deep-sea ports
	Internationalisation strategies	Rotterdam, Antwerp
2) Port operations		
Quality of input	Skills mapping and matching	New York/New Jersey
	Training and education	Singapore
	Social dialogue	Antwerp
	Upgrading equipment	Hamburg
	Land availability	
Quality of organisation	Port planning	Rotterdam
	Port information systems	Valencia
	Competition	Most large ports
	Co-ordination between ports	Copenhagen/Malmö
3) Hinterland	Links port with other transport modes	Rotterdam
	Dry ports and extended gates	Gothenburg
	Freight corridors	Betuwe-line
4) Local goodwill	Port centres	Genoa
	Port education	Long Beach
	Maritime museums	Antwerp
	Port events	Rotterdam
	Information and social media	Incheon
	Public access to port	Hamburg
	Other goodwill projects	Valparaíso

Source: OECD (2014), *The Competitiveness of Global Port-Cities*, http://dx.doi.org/10.1787/9789264205277-en.

Hai Phong's port is improving its maritime connectivity

The Port of Hai Phong is introducing deep structural changes, under the guidance of the Viet Nam Seaport Master Plan 2020, prepared by the Ministry of Transport (Box 3.1). The plan aims to modernise Viet Nam's ports and to bring them up to international operating standards. Since 2010, USD 4.5 billion has been spent by the Vietnamese government and foreign investors on renovating the ports. The rationale behind this investment is to increase the country's share of sea-based trade in the region. As previously noted, Viet Nam's share of global trade is still modest, and it lags behind countries such as Malaysia and Indonesia. The main problem identified by the central government is the lack of size and depth of Viet Nam's ports, which is limiting the economic opportunities Viet Nam could enjoy thanks to its strategic location.

Most goods exported from Viet Nam to major markets are currently trans-shipped elsewhere in Southeast Asia onto larger vessels, given the poor nautical access of Viet Nam's ports and the fact that they cannot host large vessels. The need for trans-shipment creates additional shipping and handling costs for businesses, eliminating some of the comparative advantage Viet Nam offers to investors and manufacturers (mainly its cheap labour). This is estimated to increase shipping costs by 28% in Viet Nam (Jensen, 2009). Higher maritime transport costs are related to lower external trade volumes: doubling of maritime transport costs between a given country pair is associated with a decline of 66-80% in the value of imports and a decrease in trade volume of 26-28% (Korinek and Sourdin, 2009). This is all the more important because the size of vessels is constantly increasing: in recent decades, ships have rapidly become bigger and deeper. The draft of the largest container ships is now approximately 14.5 metres, deeper than most ports can accommodate. Port depth thus becomes a competitive advantage that attracts the largest ships, and a challenge for estuary ports with no direct deep-sea access (OECD, 2014).

Box 3.1. **Viet Nam Seaport Master Plan 2020 (With Orientations Toward 2030)**

The Viet Nam Seaport Master Plan 2020, With Orientations Toward 2030, was approved by the prime minister in 2009. Its goal is to bring Viet Nam's ports up to international operating standards and to increase their competitiveness. In 2011, the Ministry of Transport prepared and passed the detailed master plan for the seaports in north Viet Nam (Group 1), i.e. those located in Hai Phong, Quang Ninh, Thai Binh, Nam Dinh and Ninh Binh provinces. The general objectives of the Group 1 Seaport Master Plan are:

- to develop seaports in north Viet Nam in a synchronous and effective manner
- to concentrate resources on developing the international gateway seaport of Hai Phong
- to develop national and regional major general seaports (Hai Phong and Hon Gai) in a sustainable, intensive and effective manner, to reduce cargo congestion and generate the momentum to develop seaport urban centres and coastal industrial parks and economic zones, promoting economic development for the whole region. The specific targets of the master plan are:

    (i) to ensure a port throughput in the planning periods of: 112 million-125 million tonnes per year by 2015; 146 million-176 million tonnes per year by 2020; and 320 million tonnes per year by 2030

    (ii) to assemble the resources to complete investment in developing the Hai Phong international gateway seaport in the Lach Huyen wharf area, to accommodate ships of up to 100 000 deadweight tonnage (DWT), facilitating the transport of Viet Nam's imports and exports on long-range shipping routes and attracting international cargo shipped through the region

    (iii) to renovate and invest intensively in raising the capacity of the ports of Hai Phong and Hon Gai and other general wharves, to exploit existing infrastructure and meet the need for handling seaborne cargoes in each area.

The Master Plan includes detailed plans to develop and renovate each seaport. Its implementation will largely be the responsibility of the Ministry of Transport and the Viet Nam Maritime Administration.

Source: Ministry of Transport (2011), *Viet Nam Seaport Master Plan 2020 - With Orientations Toward 2030.*

The Seaport Master Plan designates Hai Phong as the centre of seaport strategies in northern Viet Nam, and its port system is undergoing the most significant changes. An international gateway port and access channel is being developed off the coast at Lach Huyen, and will be able to accommodate ships of up to 100 000 DWT, or 8 000 TEUs, by 2020 (the old port system is only able to accommodate ships of up to 40 000 DWT). It is partly being developed through a public-private partnership investment by Japan's Molnykit Company and Viet Nam's state-owned enterprise Saigon Newport Corporation (VIR, 2013).[7] The new international port is being built in two stages, the first scheduled to be completed in 2016, and the second in 2020. Other terminals of the port system will receive funds to maintain their effective operations but will not be further developed, as they are located along the Cam Ca River within the Hai Phong inner city. In the long term, some of these terminals will be converted into waterfronts, a process observed in many OECD cities. The Master Plan was undertaken and financed by the Ministry of Transport and the Viet Nam Maritime Administration and has been supported by foreign investors (e.g. NYK).

These transformations of the port system of Hai Phong will significantly benefit its **maritime connectivity**, an essential element of competitiveness. With a depth of 14 metres, and an ability to accommodate larger ships, the new international port of Lach Huyen has the potential to become a trans-shipment hub, giving foreign businesses further incentives to use Viet Nam as a second regional base to offset the cost of investing in China, or to avoid using more expensive trans-shipment points such as Singapore and Hong Kong (Jensen, 2009). It is estimated that in a country that doubles its centrality in liner shipping networks, a significant increase in direct liner services to a wider range of countries can reduce transport costs by up to 15.4% (Wilmsmeier and Sanchez, 2009). A survey estimated that transport costs from the Port of Hai Phong to other ports in the world are higher than costs from the port of Ho Chi Minh City, mainly due to high cost of trans-shipment and the use of less efficient vessels of less than 1 200 TEUs (JETRO, 2012; Blancas et al., 2014). As unit costs per TEU of larger ships are significantly lower than for smaller ships, the new port could result in savings of up to 60 percent for ocean carriers, or of USD 100-200 per TEU (Blancas et al., 2014). By reducing maritime costs for businesses trading in the region directly connected to the Port of Hai Phong, the new international port is expected to increase the value of imports and the volume of trade, and to reduce the connectivity problems of Viet Nam's ports.

The central government would be ill-advised to reduce the depth of the new port at Lach Huyen to cut costs. The Port of Hai Phong would lose the opportunity to become a trans-shipment hub, forcing it to remain a feeder port, with negative consequences for its competitiveness. It will also be far more costly to adjust the depth of the port once the construction is completed, since the quay structure will need to be reinforced for additional dredging work. As a matter of fact, given the massive explosion in ship size in the recent years, even the depth and capacity of the new international port will not be able to accommodate the largest ships which in East Asia or the US West Coast are now much bigger than 100 000 DWT. This mainly owes to the fact that the Viet Nam Seaport Master Plan was designed in 2009, and that the size of ships has surged meanwhile.

A competitive port is a port that is chosen more regularly than other ports, facilitating the growth of its market share (OECD, 2014). The creation of the new international port is also an opportunity to tap into the growth of the world container fleet. International container shipping lines have difficulty filling their ships with cargo, since the balance between supply and demand for shipping tonnage since 2007 has resulted in massive overcapacity (Figure 3.5). This is expected to continue for the foreseeable future. Shipping lines are therefore desperately looking for cargo. Hai Phong's port has the potential for strong growth and should thus become of great interest to the international shipping community and shipping lines.

Figure 3.5. **World container fleet and containerised seaborne trade**

(1996-2015)

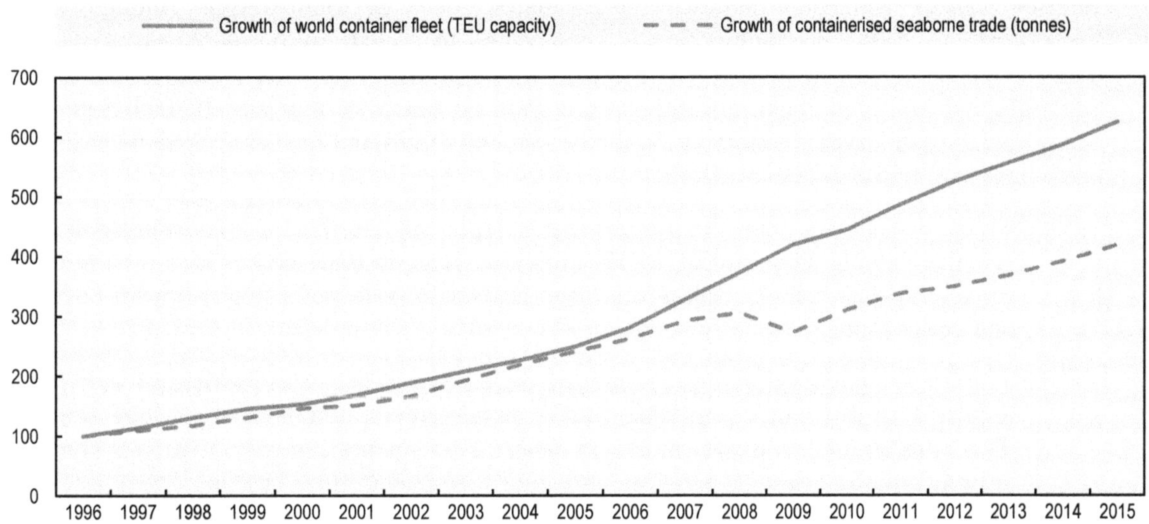

Source: OECD/ITF (2015a), "The Impact of Mega-Ships", *International Transport Forum Policy Papers*, No. 10, http://dx.doi.org/10.1787/5jlwvzcm3j9v-en.

To enhance its maritime connectivity, the Port of Hai Phong should build co-operation with other ports in the world, to benefit from consultancy and training. Such co-operation already exists: some of these arrangements are based on links from the past, e.g. partnerships of the Port of Rouen with West-African ports (Merk et al., 2011), but such partnerships could be transformed into an opportunity to attract traffic from these regions. Ports must now provide a range of incentives to shippers and operators to attract trade volume, and building co-operative relationships with other ports is one way to do this (OECD, 2014).

Enhancing the quality of hinterland connections

Several obstacles to competitiveness need to be addressed, however, in particular, the low quality of the Port of Hai Phong's **hinterland connections**. Transport infrastructure is not meeting the needs of the fast-growing cargo throughput, and containers in Hai Phong face congestion on the main roads linking the city to Ha Noi. This is principally the consequence of the heavy reliance on trucks (more than 8 000), which account for roughly 90% of the cargo transport modal split, while railways and waterways carry the remaining 10%. By comparison, the cargo transport modal split in the Rotterdam is 56% in truck transport, 10% rail and 34% by inland water. It is important to note that diesel trucks are large contributors to CO_2 and PM emissions across Asia (Clean Air Asia, 2012). The quality of roads is also an issue for the efficient conveyance of cargo. It typically takes four and a half hours for truck containers to travel from Ha Noi to the port terminals in Hai Phong, two hours more than during non-congested periods. Congestion is also typical in the terminal areas (Blancas et al., 2014).

The port system in and around Hai Phong lacks efficient multimodal freight options. The inadequacy of the railway system is the main issue. Viet Nam has only eight rail lines and around 2 500 kilometres of track. Average passenger train loads are relatively high, but freight loads are low, owing to weak infrastructure and operating plans. In addition, rail systems are diesel-based, using relatively inefficient engines (Darko, Mank Hai and Whitley, 2012). Rail only handled an estimated 1.9% of total freight volumes in Viet Nam in 2008, a figure that is projected to reach only 4.2% in 2030 (Blancas et al., 2014). Waterways of the Red River can only be used to transport bulk materials, and not containers, and there have been no plans to improve channel depths and ports along the river (Blancas et al., 2014). Several additional factors contribute to the inefficiency of hinterland connections:

- Figure 3.4 shows that the import/export split of the Port of Hai Phong has remained stable, at a ratio of 70/30. This imbalance in trade makes it necessary to move empty containers, taking up large land areas and increasing transport costs and congestion.
- The fragmentation of terminals and the dispersion of factories (while not strictly a port activity) can complicate traffic management and increase congestion. The number of terminals makes it difficult to set up a good railway system serving all the sites: at present, only one terminal (Chua Ve) has an on-dock railway station.
- Generally speaking, road infrastructure tends to be poor in Viet Nam, and the logistics system is inefficient (Maritime Insights, 2014). Dinh Vu road connections to Hai Phong City and NH5 are particularly bad, with high congestion, partly due to unpaved roads (Blancas et al., 2014).

The poor quality of hinterland connections limits the competitiveness of the Port of Hai Phong. Highway congestion resulted in additional trucking costs of an estimated USD 75.4 million in 2012, a figure expected to increase to USD 95.8 million in 2015 and to USD 144.9 million in 2020 (Blancas et al., 2014). The large share of freight transport by truck between the port and its hinterland adds to road traffic, at the expense not only of port stakeholders but the city as a whole. The low quality of hinterland connections may result in inefficiency of trade, as well as lower trade volume and economic profits. One major determinant of the relation between transport and trade is time. Each additional day in transit reduces trade volumes by 1%, leads to an increase in the freight rate of USD 56 and adds 0.8% on average to the value of manufactured goods (Hummels, 2001; Djankov, Freund and Pham, 2006). A 10% increase in time reduces bilateral trade volumes by 5-8% (Hausmann, Lee and Subramanian, 2005) and leads to a reduction in trade value of 5-25% (Nordas, Pinali and Geloso Grosso, 2006). The over-reliance on trucks rather than rail, which is less expensive, has particularly negative effects on shipping products over long distances, such as to Ha Noi and southern China (Blancas et al., 2014). In addition, congestion on urban road networks due to increased cargo throughput can negatively impact the port itself. It is widely acknowledged that port activities and transport network operations cannot function independently of each other. Inefficiency in either one will forcibly and negatively impact the other, which indicates how tightly inland networks and seaports are connected (Notteboom and Rodrigue, 2008).

The growth of cargo volumes and container throughput in Hai Phong will increase pressure on existing transport networks and calls for an urgent policy response. One of the flagship projects of the Viet Nam Seaport Master Plan is the development of an expressway connecting Ha Noi to the new international gateway port of Lach Huyen, the objective being to alleviate congestion in the city and to increase efficiency of cargo transport. The expressway will include a railway connection for cargo transport. While this project may help improve the quality of hinterland connections, the growth of container throughput and of economic activity around the port suggests that it may not be sufficient in the long term. In addition, this will improve the connection with Ha Noi but not with other provinces. Hai Phong's Green Growth Promotion Plan acknowledges the need for better railway connections, but no other concrete plan has been formulated to achieve this. Several policies should be pursued by the city and the central government to improve hinterland connections, in particular:

- A **multimodal corridor investment plan** should be set up in co-operation with the Ministry of Transport, key private sector freight stakeholders and port operators. This should include the development of the railway in Hai Phong, the modernisation of waterways and the creation of a large logistics park to accommodate multi-modal freight options. Such a park should be located along the NH5 to reduce cargo-handling costs (Blancas et al., 2014). It will be important to involve a broad range of stakeholders. Transport infrastructure planning in Viet Nam tends to be undertaken in silos, resulting in inconsistent planning and mismatched timing, and with minimal involvement of key freight stakeholders (Blancas et al., 2014).

- The current railway development between Hai Phong and Ha Noi should bring maximum benefits: in particular, the central government should co-ordinate this with the creation of dry-ports near Ha Noi. A rail terminal fairly close to Ha Noi that loads and unloads the train for further distribution of the cargo and containers by truck to their final destinations should be built to make sure the new railway is integrated into an efficient transport system.

- The modal split for freight should be regularly measured by the City of Hai Phong and the port authorities, to track progress towards more diverse, greener hinterland connections.

- The City of Hai Phong and the port authorities should explore options to create a terminal appointment system. The objective would be to reduce road congestion at port terminals by giving preferential treatment to trucks that schedule an appointment. The Gate Entry Management (GEM) system in the port of New Orleans and the Web Access system of the Georgia Ports Authority (GPA) have won recognition in this respect. Both are web-based applications that allow dispatchers to schedule appointments and provide information for pre-clearance prior to trucks' arrival at the terminal. Web Access allows customers 24-hour access to updated data on container shipments. These applications have improved traffic flow, increased terminal throughput and improved productivity for trucking companies and terminal operators, with a reduction of truck turnaround times by 30% on average in the case of GPA (OECD, 2014). Dedicated appointment lanes will need to be created at port terminals, if these measures are to operate with maximum efficiency.

- Extended gate hours could also be a successful policy for alleviating congestion. They would make possible staggering truck arrival times at port terminals throughout the day. The PierPASS programme in the port of Los Angeles-Long Beach, which includes a Traffic Mitigation Fee (TMF) for drayage transactions made during peak hours, with exemptions for off-peak hours, is a good example of such a system. It reduced daytime truck arrivals from 90% to 66% within a few months after it was introduced (Cambridge Systematics, 2009) and reduced daytime traffic on a nearby freeway by 13% (OECD, 2014).
- Fiscal incentives could be used to increase the modal split of rail and inland water transport. For instance, options to create a truck tolling system could be explored, similar to the LKW-Maut system in Germany (OECD, 2014). However, this would require alternative transport modes, which might not offer an ideal solution immediately. Another option would be to reduce port fees or handling fees for cargo trans-shipped by barge or railway.

Boosting the efficiency of port operations and creating port labour skills programmes

The efficiency of port operations is a critical element of competitiveness. It is estimated that a 10% increase in port congestion leads to a 0.7% increase in maritime transport costs (Abe and Wilson, 2009). This is related to the quality of port logistics services. Inefficient trucking services leave longer stand times on the dockside and costly inventory accumulation, and reduce export volumes, leading to less frequent shipping services (OECD, 2014). Performance indicators to measure port operations include utilisation rates of cranes, berths, yards, gates and gangs (in 20-foot equivalent units, or TEUs/year per crane, vessels/year per berth, TEUs/per year per hectare and containers/hour per lane) and productivity (moves per crane-hour, vessel service time, truck time in terminal and number of gang moves per man-hour) (OECD, 2014). One good indicator is vessel turnaround times: the average time a vessel stays in a port before departing. For ship owners and shipping lines, this is a relevant performance indicator, since port times that are longer than expected incur additional costs. This is available in detailed vessel movement data, as collected by Lloyd's Marine Intelligence Unit (LMIU). Data show that the Port of Hai Phong is relatively time inefficient, with an average container ship turnaround time of 1.66 days, higher than the turnaround time in the main Southeast Asian container ports (Figure 3.6). By comparison, the average ship turnaround time for ports in Japan is half a day, 0.7 days in South Korea, 0.8 days in China and one day in Malaysia. The average turn-around time in global container ports was 1.03 days in 2014.

A more precise indicator of handling efficiency is the berth productivity at ports. Berth productivity indicates the number of box moves per hour that a ship is at berth. This can be calculated by relating ship turnaround times to TEU volumes in the same period. Such calculations show that berth productivity in Hai Phong is poor, with less than 20 moves per hour. By comparison, the ports of Ho Chi Minh City and Singapore achieve around 70 moves per hour (Figure 3.7).

Figure 3.6. **Average container ship turnaround times in days in the main Southeast Asian ports**

(2014)

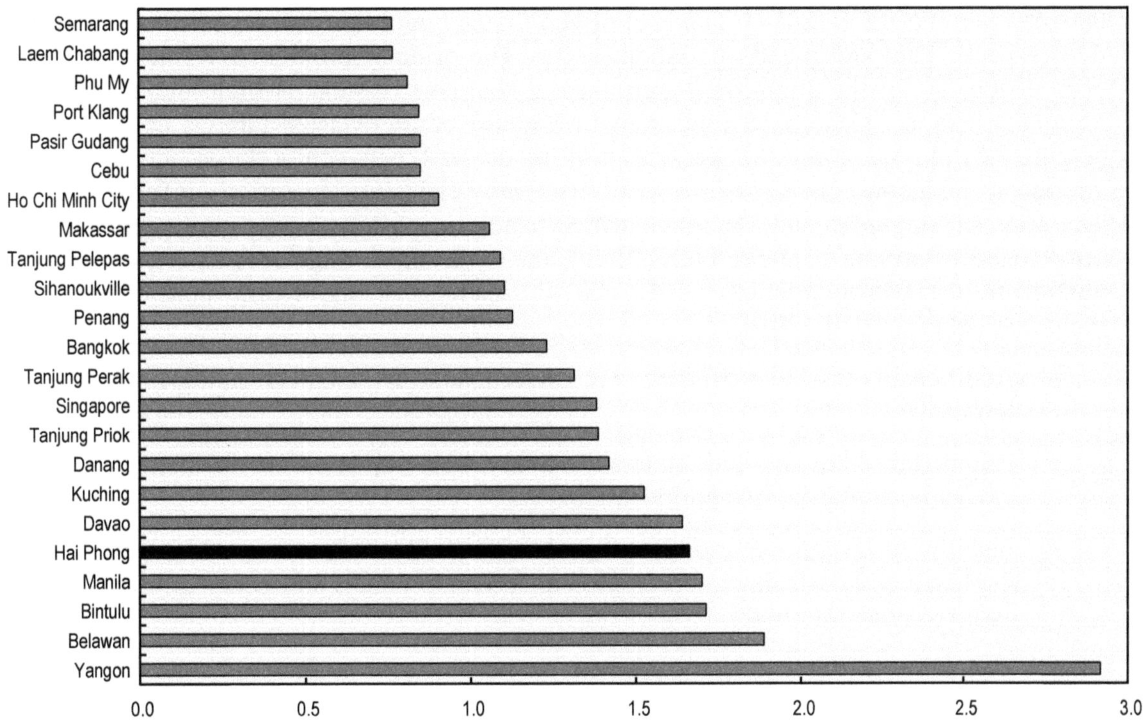

Source: ITF/OECD calculations based on Lloyd's Intelligence Unit culled from OECD/ITF (2015a), "The Impact of Mega-Ships", *International Transport Forum Policy Papers*, No. 10, http://dx.doi.org/10.1787/5jlwvzcm3j9v-en.

Figure 3.7. **Berth productivity in selected Southeast Asian ports**

(2014)

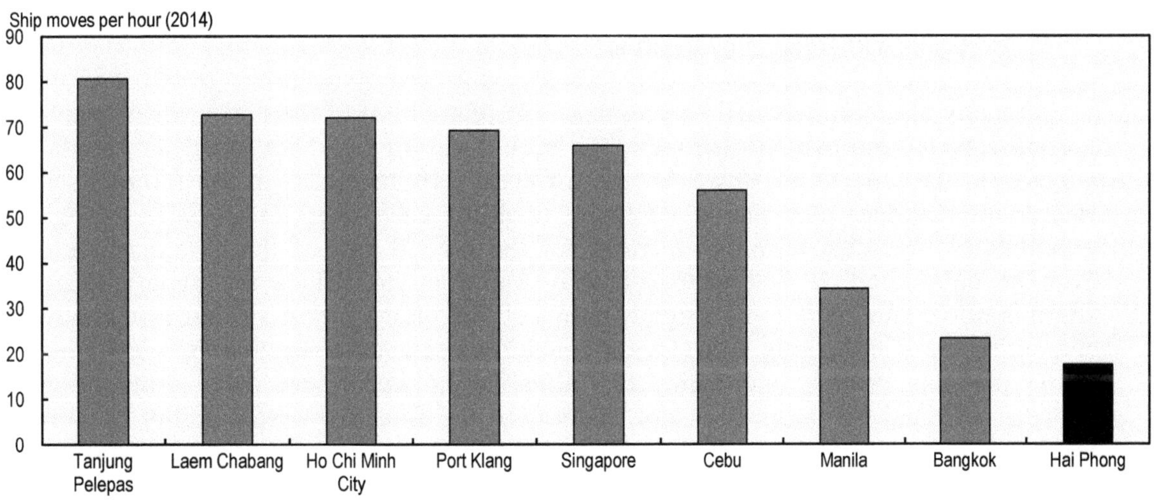

Source: ITF/OECD calculations based on Lloyd's Intelligence Unit culled from OECD/ITF (2015a), "The Impact of Mega-Ships", *International Transport Forum Policy Papers*, No. 10, http://dx.doi.org/10.1787/5jlwvzcm3j9v-en.

The efficiency of port operations depends on a number of factors, in particular the quality of inputs (e.g. labour, equipment, land), and the quality of organisations and institutions (port planning, port information systems, competition, co-ordination between ports) (OECD, 2014). The Port of Hai Phong could improve its performance in two respects: land and quality of labour skills. The primary problem for Hai Phong's port efficiency is the high fragmentation of the port system, which includes 11 terminals – soon rising to 15, once the first phase of the Master Plan is completed (Figure 3.8). In addition, the present port is divided between a large number of operators, reducing the possibility of running efficient operations (Table 3.2). The ports of Singapore and Shanghai, which handle much larger cargo volumes, have only 5 and 8 terminals, respectively. Viet Nam is the most fragmented port terminal market in the world (Blancas et al., 2014), leading to significant inefficiencies, including (Blancas et al., 2014):

- Additional land-side infrastructure investments to connect multiple marine terminals.
- Additional dredging and channel maintenance expenses to facilitate vessel navigation to the various port locations.
- Inability to leverage economies of scale at individual terminals, with significant duplication of costs due to congestion at certain terminals and under-utilisation of assets in other terminals.
- Difficulty facilitating trans-shipment volumes, with a lack of inter-terminal connections.
- Dispersion of cargo volumes, prohibiting the development of hub terminals with centralised logistics infrastructure. The system results in the fragmentation of downstream services, including inland container depots, empty container depots, bonded and non-bonded warehouses, trucking companies, and customs clearance facilities.
- Additional costs to cargo owners and ocean carriers arising from the need to maintain multiple cargo drop-off and pick-up points, or vessels making multiple port calls within the same location.

As many old terminals will be progressively phased out of the port system, major investments to improve infrastructure in each location may not yield substantial benefits in the long term. However, the different terminal areas in the future Lach Huyen international port should be large, to allow each operator to cater to high productivity. A global terminal operator for the second part of this terminal could even be considered. It will be particularly important to reduce the geographical, administrative and operational fragmentation of Hai Phong's port system to increase competitiveness in the long term. Quality should be selected over quantity, through a sound planning process and policy co-ordination, to make sure price competition between terminal operators does not undermine the financial sustainability of port facilities. This has been true of the port of Ho Chi Minh City and could be faced in Hai Phong with the development of new port infrastructure and terminals (Blancas et al., 2014).

Figure 3.8. **Hai Phong's port system**

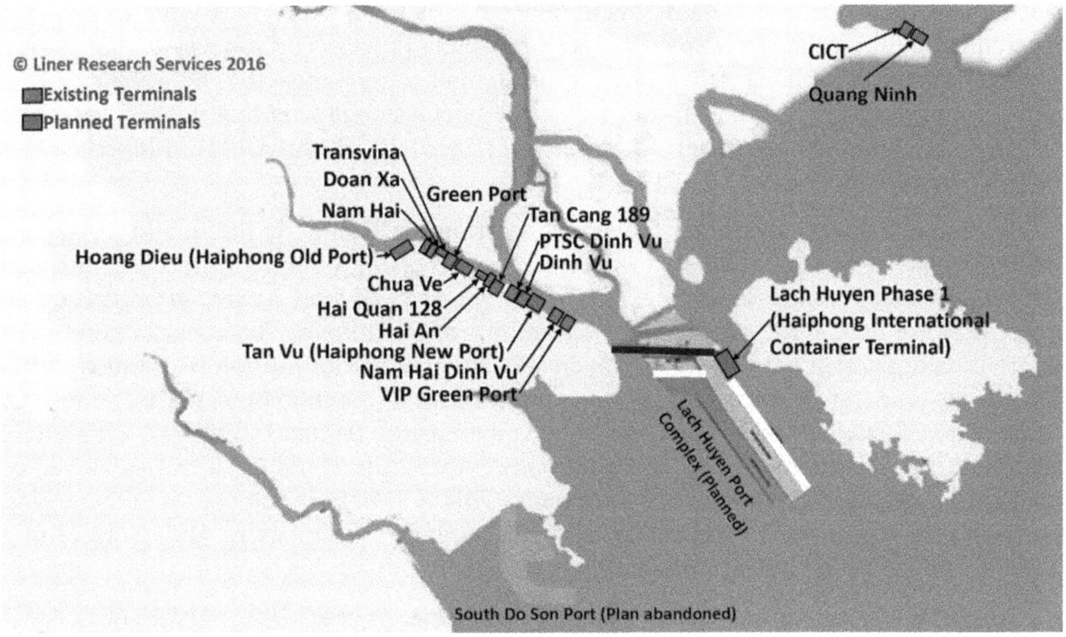

Source: Liner Research Services (2016).

Table 3.2. **Terminals and operators of Hai Phong Port**

Area	Name of terminal	Operator
Hai Phong (city centre)	Hoang Dieu	Port of Hai Phong Joint Stock Company (JSC)
	Chua Ve	Port of Hai Phong JSC
	Bach Dang	Port of Hai Phong JSC
	Doan Xa	Doan Xa Port
	Transvina	Transvina
	Green Port	Viet Nam Container Shipping JSC
	Nam Hai	Gemadept
Dinh Vu Industrial Zone (at the mouth of the estuary)	Tan Vu	Port of Hai Phong JSC
	Dinh Vu	Dinh Vu Port JSC
	PTSC	Petroleum Technical Services Corporation (Vietnamese)
	SNP 184	Saigon Newport Corporation (SNP)
	Hai An	Marina Ha Noi
	Nam Hai Dinh Vu	Gemadept/VIPCO
	Vinashin DV	Vinalines
Lach Huyen	Phase 1	Molnykit/Vinalines
	Phase 2	TBD

Source: Blancas, L.C., J. Isbell, M. Isbell, H.J. Tan and W. Tao (2014), *Efficient Logistics, a Key to Viet Nam's Competitiveness*, World Bank, Washington D.C, http://documents.worldbank.org/curated/en/646871468132885170/pdf/830310PUB0978100Box379862B00PUBLIC0.pdf.

The new deep-water port should be run on the landlord port authority model. This would mean that the publicly owned port authority owns the infrastructure and assumes responsibility for the long-term development of the port, including dredging, etc. Private terminal operators are responsible for supra-structure, including ships to shore cranes and handling equipment. The port authority then concludes concession agreements with each terminal operator for approximately 25 years, setting the general rules and targets for the operator, including the concession fee. This would allow for foreign capital investments, expertise and purchasing power. The concession fee should be both fixed and variable depending on throughput to recognise that both the operator and the port authority is working towards growth and recognising that the cost structure for the port authority is rather fixed. A 70 (fixed) to 30 (variable) split could for instance be considered.

A second critical determinant of port operations efficiency is the quality of labour skills. A lack of qualified labour to operate the increasingly modern infrastructure and services of the Port of Hai Phong has been reported, as well as low productivity in general. More than 3 400 people work in the Port of Hai Phong. The construction of the international gateway port of Lach Huyen will involve modernisation of infrastructure, increasing use of information and communications technology (ICT), and the competition between ports will create an increasing demand for port efficiency. Matching the labour pool with the port's need for skills could be a critical lever of competitiveness for the port in the long term. This will help it compete with emerging ports in the region, whereas global port cities such as Singapore and Hong Kong already benefit from highly productive labour.

Many maritime clusters now feature partnerships between universities, local government and maritime firms. These help to better match the local labour pool with the maritime cluster in three main ways: they result in new degrees and certifications that enable local students to develop skills needed by the maritime cluster and they provide access to apprenticeships and internships with participating maritime firms. They also often provide scholarships and grants for maritime education programmes, which extend the labour pool by providing greater access to education, and include the added benefit of attracting international talent (OECD, 2014). In the Rotterdam maritime cluster, for example, the Deltalinqs association has created the Maintenance College, in partnership with Albeda College and the Process College, in partnership with ROC Zadkine and the Shipping and Transport College. The curricula of such programmes are conceived in close collaboration with the maritime cluster, and respond directly to its labour demands (OECD, 2014). The Maritime and Port Authority (MPA) of Singapore offers a host of scholarships in the maritime field, which often lead to careers within the maritime cluster. Under the Tripartite Maritime Scholarship Scheme, for example, talented high school graduates are granted scholarships of up to Singapore dollars (SGD) 50 000 to complete the diploma in Nautical Studies or in Marine Engineering at the Singapore Maritime Academy or Singapore Polytechnic (OECD, 2014).

The City of Hai Phong and the port authorities should replicate such mechanisms. Creating a special college is not necessary, given that the port already benefits from the Viet Nam Maritime University located in the city. This offers a range of specialised education, with 34 bachelor degrees, 11 masters and 8 doctoral programmes. Its Vocational College develops specific skills, with a strong focus on shipbuilding and ship crew training. Around 12 000 to 15 000 students are enrolled in this programme, including 2 000 in a special international co-operation section.[8] The vocational college policy has been pursued by the local government to develop specific skills in different sectors (e.g. the Vocational College for Tourism and Services). While one of the objectives of the Vocational College of the Maritime University is to develop international co-operation, more could be achieved. Danida, the Danish development agency, is the only international partner at present, developing welding skills.

It is particularly important to involve international maritime firms in this programme, to accommodate the needs that will arise in internationalising the port, rather than simply those of Vinashin, a local state-owned shipbuilding firm.

3.3. Greening the activities of the Port of Hai Phong

Historically, generating economic welfare has been the principal objective of seaports. More recently, however, contribution to sustainable value added and low environmental impact have become new requirements, and ports are increasingly developing green portfolio analysis, with measurement of environmental impacts and calculations of the potential benefits of shifting cargo to environmentally friendly modes (OECD, 2011). While the economic benefits of port activities are widely shared with the hinterland, the environmental impact is mostly borne locally. Ports have a variety of environmental impacts, related to shipping activity, activity on the port land and the environmental impact of hinterland transport to and from ports. The main impact involves air emissions, water quality, soil, waste, biodiversity and noise. These environmental side effects can have severe consequences for the health of a port-city's population, especially on less affluent neighbourhoods (OECD, 2014). Port activity can account for a large share of a city's overall environmental impact, as is the case in Hong Kong and Los Angeles/Long Beach (Table 3.3). Greening port activity can significantly improve a city's environment and well-being.

Table 3.3. **Shipping-related emissions as a share of total city emissions**

Port	SO_2 (%)	NO_x (%)	PM_{10} (%)
Hong Kong	54	33	n.a.
Shanghai	7	10	n.a.
Los Angeles/Long Beach	45	9	n.a.
Rotterdam	n.a.	13-25	10-15

Source: OECD (2014), *The Competitiveness of Global Port-Cities*, http://dx.doi.org/10.1787/9789264205277-en.

No concrete environmental action plan for port activities in Hai Phong appears to have been drawn up, although some environmental issues have been reported, as described below. Commitments from some port operators have been formulated, in particular Hai Phong Port Joint Stock Company, in terms of environmental impact assessment, oil pollution control, hazardous waste management and sewage treatment, but further investments of this kind are needed. In addition, such commitments are not made uniformly across port terminals.[9] Hai Phong needs to address the following issues to realise its "green port city" vision.

Air pollution, low seawater quality and loss of biodiversity are primary environmental concerns

Several studies have reported negative environmental impacts of the Port of Hai Phong. The following elements are of greatest concern:

- Maritime and inland cargo traffic emit pollutants such as CO, SO_2, NO_x, and particulate matter (Tran et al., 2013). Most key industries in Hai Phong, including shipbuilding, are still highly energy-intensive and result in high air emissions. Levels of PM_{10} in Hoang Dieu, Chua Ve and Tan Vu terminals are on average 142 $\mu g/m^3$, 136 $\mu g/m^3$ and 141 $\mu g/m^3$, respectively. Dust pollution is also a negative consequence of port activities and development. This is particularly intense near the Tan Vu terminal, due to construction work and road traffic. Dust particles can penetrate the human respiratory tract and exacerbate respiratory conditions such as asthma (Fortescue, 2011).

- Research also suggests that seawater quality has been decreasing: port activities in Hai Phong result in 3 000 to 5 000 tonnes of waste oil annually, and only 900 to 1 000 (20-30%) are collected (Institute of Marine Environment and Resources, 2015).

- Loss of biodiversity is another major environmental impact. This is a consequence of deteriorating seawater quality, oil spills (especially from small gasoline-powered fishing boats) and also dredging activity. From 2001 to 2006, nearly 3 million cubic metres of materials were dredged yearly, in addition to 14.5 million cubic metres in the Lach Huyen and Ha Nam channels in 2004-05. Sediment appears to have negatively affected the ecosystems in the vicinity and biodiversity in both the source and the sediment destinations. Waste from port activities, including oil, has also contributed to the disruption of local ecosystems. Mangrove ecosystems, seagrass, tidal sands, lagoons and coral reefs, in particular, have been heavily affected.[10] Many aquatic animals, such as shrimps and crabs, which are important fishing resources, are disappearing. The upgrading of the old ports and building of the new deep-sea port in Lach Huyen has also affected local biodiversity (Tran et al., 2013). Significantly, Hai Phong is located in an estuary with valuable ecosystems of mangrove forests, coral reefs, sea grass and aquaculture areas, and the new international port is adjacent to the Cat Ba biosphere reserve area.

- Solid waste from shipping activities is another growing problem for the Port of Hai Phong. Illegal waste is found in containers disposed of by foreign industrial businesses that do not want to pay for the high cost of treating and disposing of such waste, and hazardous waste in particular. It was recently estimated that the port houses around 5 000 containers containing several thousand tonnes of illegal waste, including plastic waste, electronic parts and used rubber tubes.[11]

These impacts are felt locally, and must be addressed by any green growth strategy, especially given that the rapid growth of the port may exacerbate these problems. So far, this has not been appropriately addressed: the Hai Phong Green Growth Promotion Plan, in particular, makes almost no mention of port activities, including their environmental impact.

The need for an accurate port environmental monitoring system

The first issue to address is the **lack of an accurate environmental monitoring system for the port's activities** (including its impact on local ecosystems). Most of the issues mentioned above have not been thoroughly measured, precluding concrete action and consensus among the diverse stakeholders on what is at stake. Despite recent commitments from Hai Phong Port JSC to carry out environmental monitoring and impact assessment, no accurate time-series data on waste oil generation is apparently available. Neither are recent data on air quality in port terminals, and a clear understanding of the impact of the port on biodiversity. In addition, no data is available on other traditional environmental externalities of port activities, such as GHG emissions and the generation of solid waste. One of the strategies of the Green Growth Promotion Plan is to develop environmental monitoring systems in the city, but none of the three planned stations will be located in the port.

The City of Hai Phong and port authorities should thus build environmental impact inventories in each critical area of concern. One option would be to use the air emissions inventory developed in the port of Los Angeles (Box 3.2), in partnership with environmental institutes and NGOs, relevant port companies and the central government, with authority in the port areas. A sustainability report presenting different environmental impact indicators should be published, as is the practice in Los Angeles, Long Beach, Houston, Vancouver, Seattle, Sydney, Auckland, Hong Kong, Gothenburg, Barcelona and Hamburg (OECD, 2014).

Box 3.2. **Port of Los Angeles: How to create an Air Emissions Inventory**

The Port of Los Angeles and the Port of Long Beach have had an Air Emissions Inventory in place since 2005 to measure port-related air pollution and inform the public about it. This inventory is part of the San Pedro Bay Clean Air Action Plan (CAAP), designed to reduce air emissions and the health risks associated with air pollution (see Box 3.3). The 2005 Inventory of Air Emissions serves as the baseline for measuring progress on this action plan. The development of the air emissions inventories was co-ordinated with the US Environmental Protection Agency (EPA), the California Air Resources Board (CARB) and the South Coast Air Quality Management District (SCAQMD). Port tenants and shipping lines also play an essential role in providing accurate activity and operation information. The activity and operational data collected is then used to estimate emissions for each of the various source categories, consistent with the latest estimating methodologies agreed upon by the port and participating regulatory agencies. All the detailed annual inventory reports are available to the public on the port websites.

The inventories evaluate emissions from five port-related mobile source categories: ocean-going vessels (OGVs), harbour craft, off-road cargo handling equipment (CHE), rail locomotives (RL) and on-road heavy-duty vehicles (HDV). For each category, exhaust emissions are estimated for the following pollutants: particulate matter (PM) (10-micron, 2.5-micron), diesel particulate matter (DPM), oxides of nitrogen (NO_x), oxides of sulphur (SO_x), hydrocarbons (HC) and carbon monoxide (CO). The ports started to conduct emission estimates of greenhouse gases (GHG) from port-related operation from the 2006 Inventory, which includes carbon dioxide (CO_2), methane (CH_4) and nitrous oxide (N_2O). Using 2005 activity levels as the baseline year, the subsequent inventories also provide comparisons of main air pollutants between the baseline year and the evaluation year. In the 2011 report, the Port of Los Angeles reported a reduction in cumulative harmful emissions of 76% since 2005. Diesel particulate emissions declined by 71%, NO_x emissions by 51% and SO_x emissions by 76%.

Source: OECD (2014), *The Competitiveness of Global Port-Cities*, http://dx.doi.org/10.1787/9789264205277-en.

Developing a Port Clean Air Action Plan

In addition to environmental monitoring system for the port, specific policies could be taken for each environmental challenge, starting with air pollution. This important environmental concern has been increasingly acknowledged as an economic issue. For a city with a population of 100 000, a single tonne of $PM_{2.5}$ has social costs of approximately EUR 33 000, or social costs of EUR 495 000 for a city of several million people. The same applies to SO_2, whose costs vary from EUR 6 000 per tonne emitted to

EUR 90 000 respectively (Holland and Watkiss, 2002, cited in Castells Sanabra, Usabiaga Santamaría and Martínez De Osés, 2013). Air pollution from ports can therefore present large external costs.

The City of Hai Phong could start by developing a Port Clean Air Action Plan to tackle air pollution from port activities comprehensively, with port operators, shippers, NGOs, etc. The plan should also include measures to reduce GHG emissions from port activities. The San Pedro Bay Ports Clean Air Action Plan could offer a useful model (Box 3.3). The Action Plan should contain complementary measures to improve air quality and reduce GHG emissions, such as:

- The creation and monitoring of an Environmental Ship Index (ESI). This is the most common way for ports to incentivise the use of ships that are most environmentally acceptable, and consists in determining the environmental performance of ships with respect to air pollutants and CO_2. The aim would be to reward ships that score high on the index by offering them lower port dues (Box 3.4).
- Setting up targets for modal split and complementary fiscal incentives. As explained in the first section of this chapter, decreasing reliance on cargo transport by trucks will not only help to improve the competitiveness of the port but also reduce its environmental impact. Emissions generated by rail transport are roughly equivalent to a third of those generated by road haulage (OECD, 2014). Fiscal incentives should complement the use of rail or inland waterway transport.
- Creating a mandatory truck retirement programme. A programme of this type has been successful in reducing air emissions in Southern California. The ports of Los Angeles and Long Beach have provided more than USD 110 million in grant funding as incentives to replace trucks (OECD, 2014). In Hai Phong, this will require the financial participation of the central government.
- The Port of Hai Phong should progressively turn to the **production of cleaner and renewable energy**, in association with broader city-wide energy policies. For instance, Rotterdam's Port Vision 2030, published in 2011, is based on a strategy to link the port to its emerging sustainable energy sector (OECD, 2014). As the new port of Lach Huyen is a greenfield construction and by definition located away from where people live, wind generation could be suitable. This energy could be used for onshore power supply, as practiced in the Port of Gothenburg, Sweden (Box 3.5). This will help to reduce the burning of diesel by idling ships. Waste-to-energy plants could also be used to turn port solid waste into electricity (see Chapter 2).
- Energy efficiency in port buildings and facilities. The City of Hai Phong should extend energy efficiency measures contained in the Green Growth Promotion Plan to port buildings and facilities, perhaps as a separate targeted sub-programme.

Box 3.3. San Pedro Bay Ports Clean Air Action Plan

The San Pedro Bay Ports Clean Air Action Plan (CAAP) is a comprehensive strategy to reduce air pollution emissions from port-related cargo movement. The two San Pedro Bay ports, the largest seaport complex in North America, are also the single largest source of pollution in Southern California, according to the South Coast Air Quality Management District (SCAQMD). In 2005, the twin mega-ports of Los Angeles and Long Beach generated approximately 25% of the diesel pollution in the region (O'Brien, 2004). The CAAP aims to address the problem of the ports' growing operations and their increasing environmental impact. Its goal was to dramatically reduce emissions and associated health risks for the region without disrupting the continuous port development. The plan was first approved in 2006 and updated in 2010. Near-term plans through 2014 and long-term goals include reducing port-related emissions by 59% for NOx, 93% for SOx and 77% for diesel particulate matter by 2023 and meeting standards to lower the residential cancer risk in the port area from diesel particulates. Under the plan, the twin ports have developed annual emission inventories, which are made public, to track progress in achieving CAAP standards. The CAAP uses a combination of regulations, fees, grants and incentives to the cargo industry to promote cleaner technology and operational systems, such as the Clean Truck Program (CTP), the Vessel Speed Reduction Program (VSR) and the Alternative Maritime Power Program. To support the development and demonstration of clean-air technology, the ports have also jointly created a Technology Advancement Program that has provided more than USD 9 million in funding to the industry since 2007.

The latest analysis, from 2011, shows that the two ports have substantially reduced the key air pollutants from port-related sources since 2005, including a 71% and a 75% reduction in airborne diesel particulates, respectively. Several pillar programmes have significantly contributed to reducing air pollution at the two ports, including CTP and VSR.

The CAAP marks a milestone for the port industry in mitigating the environmental impact of maritime operations. The plan was a co-operative venture, and the two ports initiated the concept and brought along industry stakeholders and agency leaders (Giuliano and Linder, 2011). The key factor in its success is the co-operation of port users, including terminal operators, truckers and shippers, as well as the support of federal, state and local regulatory bodies and local communities (Mongelluzzo, 2012). The ports were also under considerable social pressure. Community concern over the health risks of port-related diesel emissions had grown after a series of air quality studies was published on the correlation between cancer and respiratory disease rates and proximity to freight-movement corridors. Cargo volumes rose through 2004, in an expansion of capacity at the two ports, and public opposition, including a series of lawsuits, made plans for expansion difficult, if not impossible. Political pressure for increased regulatory oversight also prompted the ports to respond to public dissatisfaction over air quality. This ultimately led to the adoption of a comprehensive plan. The CAAP was portrayed as a solution for building the credibility of the ports to obtain agreements on future projects as they engaged all the key stakeholders. One study describes the CAAP as "a response to the loss of social legitimacy and to social and regulatory pressures that were restricting the ability of the ports to expand" (Giuliano and Linder, 2011). The two ports' market influence also played a role in the mitigation efforts, since their gateway location gave them more room to impose fees on the industry and generate the revenue to carry out environmental policies.

Source: O'Brien, T. (2004), "Quality of life and port operations: Challenges, successes and the future", white paper prepared for the sixth annual CITT State of the Trade and Transportation Industry Town Hall Meeting, Center for International Trade and Transportation; Giuliano, G. and A. Linder (2011), "Motivations for voluntary regulation: The Clean Air Action Plan", submitted for presentation at the 2012 Transportation Research Board Annual Meeting, http://docs.trb.org/prp/12-2628.pdf; Mongelluzzo, B. (2012), "How California's ports cleared the air", *Journal of Commerce*, 23 January, www.joc.com/portsterminals/how-californias-ports-cleared-air?page=2, (accessed 23 July 2016); OECD (2014), *The Competitiveness of Global Port-Cities*, http://dx.doi.org/10.1787/9789264205277-en.

Box 3.4. **The Environmental Ship Index**

The Environmental Ship Index (ESI) helps to determine the environmental performance of ships with respect to air pollutants and CO2. The aim is to reward ships that score high on the index by offering them lower port dues. The ESI measures a ship's emissions based on the amount of nitrogen oxide, sulphur oxide, particulate matter and greenhouse gas it releases. A voluntary system, it is open to shipping companies, ship owners and ports. It uses a formula to give ships points for their environmental performance, taking into account current international legislation, mainly promulgated by the International Maritime Organisation (IMO). At present, 1 439 ships have a valid ESI score, and 48 ports participate, including Tokyo, Rotterdam, Hamburg, Antwerp, Le Havre, Los Angeles and New York/New Jersey. The ESI was developed in the framework of the World Port Climate Initiative (WPCI), committing to reduce greenhouse gas emissions from port activity. The ESI ship database is administered by the ESI Bureau of the International Association of Ports and Harbors (IAPH).

Source: OECD (2014), *The Competitiveness of Global Port-Cities*, http://dx.doi.org/10.1787/9789264205277-en.

Box 3.5. **Onshore power supply in the Port of Gothenburg**

Since the turn of the millennium, the Port of Gothenburg in Sweden has launched an innovative policy of using onshore power supply. Vessels in dock typically use their diesel engines to meet energy needs for certain functions, such as lighting, heating and air conditioning. This use is a source of local air pollution and greenhouse gas emissions. The Port of Gothenburg was the first in the world to propose that vessels be connected to the local energy network, making it possible to shut off their engines during their stay in the port (called "cold ironing").

Since 1989, the Port of Gothenburg has provided electricity to ships calling at the port, but only with low-voltage cables that did not meet all their energy needs. At the initiative of a large paper-manufacturing company, Stora Enso, which was trying to improve its carbon footprint when transporting its products, the port began designing a more efficient system in partnership with several navigation companies and Asea Brown Boveri (ABB), a company specialising in electrical products. Operational since 2000, this newer system uses a single high-voltage cable providing 6.6 to 10KW 50Hz, which can power an entire ship from platforms on the docks. The vessels are thus able to turn off their engines, resulting in a significant reduction in both noise and carbon emissions. The Port of Gothenburg estimated that a vessel not connected to the onshore power grid emits about 25 tonnes of carbon dioxide, 520 kilogrammes of nitrogen oxides and 22 kilogrammes of particulate matter while in dock. This innovation thus benefits both the environment in terms of climate change, and the quality of life and working conditions of the population at or near the port (residents, dockworkers and ships' crews). Today, one in three ships calling at the Port of Gothenburg uses the connection for shore-side electricity, a proportion that is likely to increase. Roll-on/roll-off ships and ferries are the most frequent users of the new system, because the links they provide are back and forth, but all categories of ships may benefit from this new technology. While connecting to the grid requires vessels to invest in technology to use the new system, costs for retrofitting vessels can be offset by the likely savings in fuel.

Through this programme, the Port of Gothenburg has acquired a first-mover technology advantage in connecting the vessels to shore-side electricity. This system is also in operation in other ports, such as Antwerp (Belgium), Zeebrugge (Belgium) and Lübeck (Germany). However, a significant barrier to technology diffusion is that international electricity standards are not harmonised, with some parts of the world using 50 Hz systems and others 60 Hz systems. This complicates retrofitting vessels, although attempts are being made to align them. For its pioneering role in this technology, the Port of Gothenburg was chosen as the leader of the Working Group on onshore power supply created by the World Port Climate Initiative.

Source: OECD (2014), *The Competitiveness of Global Port-Cities*, http://dx.doi.org/10.1787/9789264205277-en.

A fixed-fee system and green bunkering programme can reduce waste and discharge of sludge

Port activities produce solid and non-solid waste, especially from oil terminals, fuel deposits and dry dock operations, which produce oily and toxic sludge. Waste also comes from other sources, such as ships (Miola et al., 2009), and, as mentioned previously, illegal dumping in containers by foreign businesses. Shipping activities are responsible for around one-fifth of global discharges of waste and residues at sea (EMSA, 2013). In Hai Phong, as noted previously, only 20% to 30% of oil waste is collected, the rest being discharged into the sea. Oil spills result from normal activities, accidents and illegal dumping practices, such as port run-off, unloading and loading of oil tankers, removal of bilge water and leakages. Although tanker accidents are thought of as an important source of water pollution, some estimates indicate that normal shipping operations are responsible for over 70% of the oil discharged into the sea from marine transport (Miola et al., 2009). Other types of waste are also often released into the sea. Also, given the number of shipyards for repair and new builds in Hai Phong, it is likely that the dredged masses are highly contaminated. Record pollution of sea water was reported in 2013 in Hai Phong.

Hai Phong already has several measures in place to limit solid waste and non-solid waste pollution. Port handling companies sign contracts with Hai Phong Solid Waste Management Enterprise to use the solid waste collection service. One unit in the company is responsible for treating waste collected from the port. Waste is directly transported to landfill sites. The fee is determined by the local government and depends on the volume of waste collected (for toxic waste, the fee depends on the type of waste). An environmental fee is levied in the port, but is only used to finance the cost of basic port cleaning. Such measures are, however, obviously insufficient, and given the growth of port activity, the need for financial and human resources capacities will only increase. The city faces the risk that environmental impact assessment will become harder and harder to enforce, and that solid waste treatment facilities will struggle to deal with increasing amounts of solid waste. The port, generally speaking, lacks consistent control of solid waste and contaminated mass. The Green Growth Promotion Plan, for instance, acknowledges the need to reduce wastewater from industries and residences but does not mention port activities.

The port should thus develop a cleaner production strategy to minimise waste, and control and manage the quality of the ships that enter the port and the process of loading goods. The single most effective policy option local and port authorities could consider is imposing on ships a **fixed-fee system for waste**. The countries in the Baltic Marine Environment Protection Commission (HELCOM) in northern Europe adopt the "no special fee" system for collecting waste and sludge generated by ships. Under this agreement, ports are not allowed to charge ships based solely on the amount of sludge and waste left in port but rather a fee for all ships calling in ports that is based on gross tonnage. The rationale for this policy is to remove the incentive for the shipping line to dump waste and sludge in the ocean. If ships are required to pay for waste and sludge regardless whether they leave it in the port or not, they have no economic reason to dump it in the water. Introducing this system in the Port of Hai Phong could thus reduce the waste, especially illegal waste, and sludge dumped in the water. It might, however, be necessary to harmonise such a policy throughout Viet Nam and even the region, to ensure that the Port of Hai Phong does not receive too much waste, if other ports base their fees on the amount of waste. The national government could play a critical role in this regard. The fee should be set high enough to avoid this problem.

One other policy option would be to create a "green bunkering programme" along the lines of the one in force in Gothenburg. This port, which handles half of Sweden's oil imports, has undertaken a range of measures to ensure that gases and oil are not inadvertently discharged into the environment during bunkering. The port introduced a stringent set of rules in 1999 covering a wide range of activities, including requiring the installation of electronic overflow alarms, the carriage of at least 50 metres of oil booms with absorptive material, and the vetting of all bunker barges by the port authority. The port has also mandated oil-spill prevention equipment for bunker installations and that all bunker operators attend training programmes to learn safe bunkering techniques. Gothenburg has argued for the expansion of green bunkering practices to the rest of Sweden, supporting a 2011 bill to require regular pressure testing in Swedish bunkers to prevent oil spills (OECD, 2014).

Creating no-discharge zones to protect local biodiversity

Port activities can impact local and global biodiversity (Table 3.4). In Hai Phong, oil discharge and sedimentation are most frequently quoted in studies as contributing to a loss of local biodiversity. Mangrove losses, in particular, are a source of concern, and local authorities have undertaken a series of measures to protect these important natural assets. Programmes and projects investing in environmental conservation in Hai Phong for the period 2012-25 include coastal mangrove protection and recovery projects with an estimated budget of around VND 85 billion.

Restoration of biodiversity is critical, and the City of Hai Phong should adopt pro-active policies and reinforce current biodiversity protection policies. No-discharge zones in and around the city could be set up, where biodiversity is particularly high and sensitive. The objective would be to forbid the discharge of certain liquid waste into the sea. Around 30 states in the United States have created such no-discharge zones. Three different mandates for protection are used: protecting aquatic habitats where pump-out facilities are available; protecting special aquatic habitats or species; or protecting drinking-water intake zones to protect human health. In March 2012, the largest no-discharge zone in the United States was created along the coastline of California (2 600 km), and should reduce the annual discharge by over 76 million litres of sewage by cargo and cruise ships calling at California's ports (OECD, 2014).

Table 3.4. **Port impacts on biodiversity**

Source	Effects	Species affected
TBT paint	Morphological change, change in population structure	Marine invertebrates
Anchoring	Sediment resuspension, reduction of capacity for photosynthesis	Marine organisms living in harbours, seagrass
Oil discharge	Genetic damage, oxidative stress, behavioural abnormalities	Marine vertebrates, birds
Gas emissions	Ocean acidification	Plankton, coral, organisms with calcification process
Chemicals	Accumulation of substances in organisms that cause disruption of the endocrine system	Predators at the top of the food chain
Waste	Eutrophication	Seagrass, fish
Debris	Death by ingesting floating plastics	Seabirds, turtles, whales
Ballast water	Introduction of invasive non-indigenous species, extinction of native species	Entire ecosystem
Noise	Problems of communication for animals, collisions	Cetaceans, marine mammals
Collisions	Death	Cetaceans, other marine vertebrates (whales, dolphins, turtles)

Source: OECD (2014), *The Competitiveness of Global Port-Cities*, http://dx.doi.org/10.1787/9789264205277-en, based on Abdulla, A. and O. Linden (eds.) (2008), "Maritime traffic effects on biodiversity in the Mediterranean Sea: Review of impacts, priority areas and mitigation measures", IUCN Centre for Mediterranean Co-operation.

Integrating green port policies into Hai Phong's Green Growth Promotion Plan and Action Plan

So far, environmental consideration of port activities has been not sufficiently provided for under current policies. First, the specific green growth challenges of the Port of Hai Phong are not included in the Hai Phong Green Growth Promotion Plan (GGPP) and Green Growth Action Plan (GGAP), although the "green port city" is the central vision of both strategic documents. The specific issues associated with the port that are obstacles to green growth in the city (e.g. multimodal transport and the environmental challenges described above) are not clearly defined and recognised in the wider city-scale green growth policies. The recommendations defined above (including the development of multimodal transport) should thus be integrated into these documents, either in a specific green port section, or incorporated into the sectoral policies of these plans (in particular in the sections concerning solid waste management, decentralised energy systems, air quality monitoring, transport and energy efficiency in manufacturing).[12] To ensure policy coherence across levels of government and increase the impact of local green growth strategies, the Viet Nam Seaport Master Plan should integrate green growth principles and be aligned with national green growth strategies. This would have a high impact, as the national government party owns most of the port terminals in Hai Phong, through the Viet Nam National Shipping Lines (Vinalines).

Designing and implementing green port strategies will also require the involvement of private port operators. The City of Hai Phong does not own the port terminals, which may also explain why specific port activities are not included in the GGPP and GGAP. It also appears that the local government is not privy to much of the data on the port systems, which is mostly owned by private operators. Green port strategies should benefit all stakeholders, and by closely involving port operators, the City of Hai Phong should be able to make its policies more inclusive, with greater impact on the whole city.

3.4 Enhancing the port's resilience to typhoons and floods

Exposure to typhoons and floods

The risk of natural disasters is a potentially critical obstacle to green growth in Hai Phong. The city has a history of storms and flooding going back centuries, and was almost completely destroyed by a typhoon in 1881 that killed 300 000 people. It experiences an average of three to four storms with heavy rains and strong winds every year, in particular in July and August, during the monsoon season. Other extreme weather events, such as tornadoes, were reported in 2008 and 2011. Among cities of over a million inhabitants, Hai Phong is among the world's top 20 coastal cities whose population is most exposed to coastal floods, with 794 000 residents exposed as of 2011. It is projected that that number will grow to more than 4.7 million by the 2070s, under a scenario taking into account climate, subsidence and socio-economic changes (see Chapter 1). While Hai Phong is not among the top 20 cities whose assets are the most exposed, it is among the top 15 cities likely to be exposed to the highest proportional increase in exposed assets by the 2070s, under the same scenario. A concrete example of a recent natural disaster in the city is the Son Tinh typhoon in 2012, which caused loss of property estimated at VND 1 000 billion (around USD 330 million, PPP), and destruction or damage of many businesses and infrastructure, including 63 wrecked ships and 8 collapsed marine management stations. This typhoon was the city's most devastating in the past ten years.

The city also suffered from similar disasters in 2005, 2008, 2009, 2010 and 2011 affecting port facilities. The 2008 storm overturned 202 containers and destroyed two ship cranes.

Several reasons can explain this increasing vulnerability:

- The City of Hai Phong lies directly on one of the most frequent paths for Pacific typhoons, which originate in and around the Philippines and reach the Asian mainland through the Gulf of Tonkin. Severe tropical cyclones are expected every 5 or 10 years, and statistical data shows that the total number of storms annually is on the rise, owing notably to climate change. These storms can directly damage infrastructure or increase the water level and risk of flooding in the city's numerous waterways. Hai Phong has the highest density of rivers of cities in northern Viet Nam.
- Another risk factor in the port-city is sea-level rise, which will affect all coastal cities of Viet Nam in the long term (Figure 1.21). The entire delta region, backed by the steep rises of the forested northern highlands to the west and north, is no more than three metres above sea level, and much of that, including Hai Phong, is one metre or less above sea level.
- In addition, the natural coastal defences such as mangrove forests and estuary wetlands have disappeared, mainly due to urban and port development.

An additional risk factor linked to floods and storms are the harmful substances, materials and critical facilities in the port, such as oil, hazardous waste in containers, sludge, etc. The intensity of climatic events and the lack of proper treatment and handling of such materials create a risk of environmental contamination during floods and storms.

While the local government can do little to prevent sea-level rise or typhoons, a number of other factors of vulnerability are closely related to development, preparedness and infrastructure and can be more concretely addressed. Building resilience must be a key strategy of the local government, with the port at the centre of such policies. Given its importance in the regional economy and its density of infrastructure, its vulnerability to typhoons and floods could have severe economic repercussions on the entire agglomeration. Port activities that can increase the city's vulnerability include unregulated or uncontrolled ship-generated waste; port operations (e.g. inadequate temporary storage and pre-treatment of storm water run-off from port terminals); and land reclamation for port development projects. Building adaptive capacities for resilient urban planning, infrastructure management and emergency response should be the overarching objective of local and port authorities.

Adaptive urban planning and infrastructure management for a resilient port-city

The local government and the port authority are currently implementing measures to build resilience of the port-city to typhoons and floods, through urban planning and infrastructure. The main past or current policies include:

- Construction of dikes and pumps along the coast or the rivers. For the period 2012-25, 14 dike/embankment renovation or construction projects of an estimated total budget of more than VND 3 000 billion were listed as part of the environmental protection programme of the City of Hai Phong.
- Construction of retention ponds in the city to retain storm water and prevent it from flowing into inhabited urban areas (including port areas).

- Creation of a system of anchorage under port cranes, to ensure their stability during a storm. For the period 2012-25, VND 65 billion has been committed to renovate river ports and seaports, including construction of safe anchorage for larger ships during storms.

The local government and the port authority, however, need to work on urban planning and adaptive infrastructure strategies to build the port's resilience, especially considering that more intense and unpredictable storms are projected as a result of climate change. The following strategies should be pursued:

- Public and port authorities should ensure that land-use and infrastructure planning at the port takes into consideration **future downscaled climate change projections** for the Hai Phong area and northern Viet Nam. They should consider the combined risks (probability of impact or damage) for both sea-level rise and storm-surge levels associated with large potential storms and typhoons (and tidal levels for the interior sections of the old port along the river). Only one project, of VND 5 billion, is now listed in the 2012-25 Environmental Conservation Programme to prepare for climate change and sea-level rise in Hai Phong.
- Port resilience strategies should include **adaptive management of infrastructure**. As part of the National Pollutant Discharge Elimination System (NPDES), most US ports require Standard Urban Stormwater Mitigation Plans (SUSMPs), indicating a range of measures for effectively dealing with storm-water run-offs. The Port of San Diego's SUSMP, for example, integrates several design concepts into any new commercial and industrial development projects larger than one acre in dock and maintenance areas of the port. Requirements under the SUSMP include the use of pervious surfaces (permeable pavements and surface structures), and the planning of land slopes and gradients to ensure drainage or retention in designated zones (OECD, 2014). Similar options should be explored in the Port of Hai Phong, and integrated as requirements for building permissions. It will be particularly useful to "make space" for water in case of heavy rainfall, on which dikes and embankments can have only a limited effect.

Building preparedness and response capacities in the port

Preparedness and response capacities also influence resilience. One remarkable measure listed in the 2012-25 Environmental Conservation Programme of the City of Hai Phong is a project with a budget estimated at VND 4 billion to strengthen communication systems to warn vessels in case of typhoons. However, no comprehensive emergency response strategy is in place, and little is known about preparedness plans for staff. The following two strategies should be pursued:

- It is critically important that the infrastructure (e.g. access roads, hazardous materials or waste storage sites, emergency energy supply systems, emergency command centre co-ordinating response with necessary equipment pre-positioned and trained personnel on site), assets (ships, crane equipment, etc.), and personnel to handle accidental releases of hazardous chemicals, other materials and waste are **controlled/contained quickly** to prevent major or irreversible damage or risks. A containment plan could be set up jointly between the City of Hai Phong and port authorities/operators, so that both parties share resources and co-ordinate emergency response in case of disaster. Early warning systems should be set up in each port terminal. These could be triggered by weather forecasts and by sensors, with gauge stations in the Cam Ca River.

- The local government and the port authority should **train emergency response staff and revise emergency response plans on an ongoing basis** to reflect the changing risks. This should be based on the experience gained in mock emergency response drills for developing effective response protocols and procedures. Periodically, the plans should be reviewed by an independent third party, to ensure that they cover the main threats that are anticipated and that they remain flexible enough to respond to unexpected threats.
- Response capacities should be reinforced to ensure the economic resilience of the port. Port companies should develop **business continuity plans** to minimise the impact of a disaster on their activities.

Creating institutional synergies to enhance the port's resilience

Institutional synergies should be created to mobilise the knowledge and resources of all stakeholders who can help build up the resilience of the port. Disaster units in each port terminal are already in place, but much broader co-operation and co-ordination is needed. In particular, the following institutional co-operation strategies could be explored:

- First, given its authority in port development and knowledge of coastal resilience issues in Viet Nam, the central government could play a leading role in supporting a local port resilience action plan. In the United States, the Office of Ocean and Coastal Resource Management of the National Oceanic and Atmospheric Administration (NOAA) has developed "Adapting to climate change: A planning guide for state coastal managers" to help federal and state coastal managers develop adaptation plans. Other plan-based strategies of this sort include data storage plans, emergency response and recovery plans. These plans often involve practices such as drills and event reconstructions, simulation of post-storm actions and storm preparations (OECD, 2014).
- Second, **PPPs should be formed with shipping, operating and other port companies** to acquire more expertise, resources and assets to prevent and manage accidental or intentional spills or releases before, during or after disasters. This would help share the risks with businesses and provide incentives for them to participate in the resilience efforts. A "safe" or "eco" port branding campaign could also offset the costs of the port resilience projects and increase the port's competitiveness and market share. The goal would be to change the port's marketing strategy. Rather than offering the cheapest labour costs or other requirements to use its ports in a "race to the bottom", its superior port facilities, services and safety record would provide an incentive for shipping companies to pay a premium price to use Hai Phong as a sea gateway to northern Viet Nam and southern China. The aim would be to attract customers looking for high-quality port facilities and services, not those looking for the cheapest deal possible.

Creating a resilience action plan to increase the port's adaptive capacity

All these measures should be regrouped into a resilience action plan jointly prepared by the central government, the local government and relevant port authorities and operators. Its objective should be to increase the port's adaptive capacity by: i) reducing exposure to risks by making projections of future potential threats and damages using downscaled global climate change models; ii) building flexible infrastructure, such as semi-permeable surfaces, in each terminal; iii) tracking the purchase of all hazardous materials in the port or on the ships, and ensuring their safe and responsible use, storage and final disposal with a "cradle-to-grave" manifest system that tracks possession of these materials at every step, to affix responsibility and liability; iv) having the equipment, training and other resources available on site to prevent or control any contingency from turning into a "man-made" disaster (equipment and trained personnel, for example, must be available during or after an emergency); and v) unencumbered access to top port management with some external review or some joint or shared responsibility for the safe operation of the port, and immediate recovery from a disaster like a typhoon.

The local government should take advantage of the ongoing partnership with Peace Wind America and the City of Seattle on disaster risk resilience to involve port authorities and private operators in building a local resilience action plan. The City of Seattle is a historical sister City of Hai Phong, and the partnership has been established within USAID's Sister Cities Disaster Preparedness Programme, which aims to increase business resilience and public-private co-ordination of disaster-risk planning. The project will conduct risk assessments, preparedness planning, hazard mapping and business continuity trainings and consultations. It is expected that more than 2 000 businesses will assist civil authorities in substantive planning and training activities through late 2016 (USAID, 2014). This offers a timely opportunity to include private port stakeholders working on port facilities, and also all businesses whose activities are port-related (e.g. shipbuilding and repair).

Main policy recommendations

- Create a **Multimodal Corridor Investment Plan** in co-operation with the Ministry of Transport, key private sector freight stakeholders and port operators, to improve **hinterland connections.**
- Support the construction of **new railways** through reductions in port dues, create a terminal appointment system and extend gate hours.
- Measure **port operations efficiency** through vessel turnaround times, and boost efficiency by creating larger terminal areas for each operator in the Lach Huyen terminal, as well as working with international maritime firms at the vocational college of the Maritime University to improve local port labour skills.
- Develop an accurate **environmental monitoring system** for port activities, and draw up a port clean air action plan, including such measures as the adoption of Environmental Ship Index (ESI), equipping terminals with onshore power supply to avoid high diesel burning by idling ships, and mandatory retirement programmes for old trucks.
- Create green bunkering projects to decrease oil spill into the water, and introduce the **"no special fee" system** for collecting waste and sludge generated by ships.
- Take into consideration future downscaled climate change projections in land-use and infrastructure planning at the port, require semi-permeable surfaces in new buildings and roads, train emergency response staff and revise the emergency response plan on an ongoing basis, to increase **the resilience of the port to typhoons and floods.**
- Integrate the competitiveness, environmental and resilience challenges and recommendations specific to the port into the broader Hai Phong Green Growth Promotion Plan and Action Plan; to increase policy coherence and impact, the national government should also integrate green growth into its Viet Nam Seaport Master Plan.

Notes

1. This does not include Lao PDR, the only landlocked country of the region.
2. See: Hai Phong Green City website article, http://Hai Phongtangtruongxanh.gov.vn/en/reports-presented-at-the-seminar-city-building-green-port-from-practical-step-by-step-approach-to-Hai Phong/(accessed 23 July 2016).
3. Growth rate for Cebu refers to 2005-13.
4. See: Hai Phong Green City website article, http://haiphongtangtruongxanh.gov.vn/en/to-haiphong-become-the-green-port-city/ (accessed 23 July 2016).
5. The Figure 15% here includes terminals from Hai Phong and Cai Lan, for the year 2011.
6. See: Viet Nam Investment Review, www.vir.com.vn/Hai Phong-to-capitalise-on-a-good-year.html (accessed 30 March 2016).

7. Vinalines, the country's state-owned port and shipping company operator, was initially designated as the developer, but does not have the capacity to raise the funds to build the new site (Blancas et al., 2014).

8. See: VMU Vocational College website, http://eng.vimaru.edu.vn/faculty-staff/vmu-vocational-college.vmu-0 (accessed 30 March 2016).

9. See: Hai Phong Green City website reports, http://Hai Phongtangtruongxanh.gov.vn/en/reports-presented-at-the-seminar-city-building-green-port-from-practical-step-by-step-approach-to-Hai Phong/(accessed 23 July 2016).

10. See: Hai Phong Green City website reports, http://Hai Phongtangtruongxanh.gov.vn/en/reports-presented-at-the-seminar-city-building-green-port-from-practical-step-by-step-approach-to-Hai Phong/(accessed 23 July 2016).

11. See: Viet Nam News article, http://vietnamnews.vn/environment/265442/banned-industrial-waste-leaking-in-via-import-loopholes.html, accessed 31 May 2016, and Thanh Nien News article, www.thanhniennews.com/society/industrial-toxins-found-in-15-containers-at-hai-phong-port-15781.html(accessed 31 May 2016).

12. The transport policies of the Hai Phong Green Growth Promotion Plan (HPGGPP) include the construction of a new railway line to Lach Huyen. No detailed action plan is in place, however, and the measure does not cover other forms of sustainable transport such as waterways, and is thus not comprehensive. Energy efficiency in port-related manufacturing, although not mentioned in this chapter due to a lack of data on air pollution from port industries, could also be integrated into the energy efficiency measures of the HPGGPP, if appropriate.

Bibliography

Abdulla, A. and O. Linden (eds.) (2008), "Maritime traffic effects on biodiversity in the Mediterranean Sea: Review of impacts, priority areas and mitigation measures", IUCN Centre for Mediterranean Co-operation, Malaga, Spain, https://portals.iucn.org/library/efiles/documents/2008-042-1.pdf.

Abe, K. and J. Wilson (2009), "Weathering the storm: Investing in port infrastructure in lower trade costs in East Asia", *World Bank Policy Research Paper*, No. 4911, World Bank, Washington DC, ftp://ftp.worldbank.org/pub/repec/SSRN/staging/4911.pdf.

Blancas, L.C., J. Isbell, M. Isbell, H.J. Tan and W. Tao (2014), *Efficient Logistics, a Key to Viet Nam's Competitiveness*, World Bank, Washington D.C, http://documents.worldbank.org/curated/en/646871468132885170/pdf/830310PUB0978100Box379862B00PUBLIC0.pdf.

Cambridge Systematics (2009), *FHWA Operations Support – Port Peak Pricing Program Evaluation*, report for Federal Highway Administration, US Department of Transportation, http://www.ops.fhwa.dot.gov/publications/fhwahop09014/fhwahop09014.pdf.

City of Hai Phong (2015), "Hai Phong City – Integration and development", Department of Foreign Affairs, presentation made on 24 June 2015, Hai Phong, Viet Nam.

Clean Air Asia (2012), "Accessing Asia: Air Pollution and Greenhouse Gas Emissions Indicators for Road Transport and Electricity", Pasig City, Philippines.

Darko, E., N. Manh Hai and S. Whitley (2015), "Mapping current incentives and investment in Viet Nam's transport sector: Informing private climate finance", Overseas Development Institute, London, https://www.odi.org/sites/odi.org.uk/files/odi-assets/publications-opinion-files/9642.pdf.

Djankov, S., C. Freund and C. Pham (2006), "Trading on Time", *World Bank Policy Research Working Paper*, No. 3909, World Bank, Washington DC, https://openknowledge.worldbank.org/bitstream/handle/10986/8674/wps3909.pdf.

Dynaliners (2015), Dynaliners Newsletter, June 2015.

EMSA (2013), "Port waste reception facilities", European Maritime Safety Agency, http://emsa.europa.eu/implementation-tasks/environment/port-waste-reception-facilities.html (accessed 25 June 2016).

Fortescue (2011), "Port facility: Dust environmental management plan", Fortescue Metals Group, East Perth, Australia, http://fmgl.com.au/media/2211/port-facility-dust-environmental-management-plan107.pdf.

Giuliano, G. and A. Linder (2011), "Motivations for voluntary regulation: The Clean Air Action Plan", submitted for presentation at the 2012 Transportation Research Board Annual Meeting, July, http://docs.trb.org/prp/12-2628.pdf.

Hausmann, W., L. Lee and U. Subramanian (2005), "Global logistics services, supply chain metrics and bilateral trade patterns", Mimeo, World Bank, Washington DC, http://elibrary.worldbank.org/doi/abs/10.1596/1813-9450-3773 (accessed July 2016).

Holland M.R. and P. Watkiss (2002), in M. Castells Sanabra, J.J. Usabiaga Santamaría and F.X. Martínez de Osés (2013), "Manoeuvering and hotelling external costs: enough for alternative energy sources?" *Maritime Policy and Management*, the flagship journal of international shipping and port research, pp. 1-19.

Hummels, D. (2001), *Time as a Trade Barrier*, Mimeo, Purdue University, West Lafayette, Ind., July, http://www.krannert.purdue.edu/faculty/hummelsd/research/time3b.pdf.

Institute of Marine Environment and Resources (2015), "Issues of integrated management for the port environment in Hai Phong toward green and blue ports", presentation made by Tran Din Lanh at the 4th OECD Workshop on Urban Green Growth in Dynamic Asia, 25-27 June 2015, Hai Phong, Viet Nam.

Jensen, A.C.M. (2009), "Viet Nam's port potential: The economic and political implications of Viet Nam's port renovation", Project 2049 Institute, Arlington, Virginia, https://project2049.net/documents/vietnams_port_potential_shipping.pdf.

JETRO (2012), *The 22nd Survey of Investment Related Costs in Asia and Oceania (FY 2011 Survey)*, Japan External Trade Organization, Overseas Research Department, Tokyo, https://www.jetro.go.jp/en/reports/survey/pdf/2012_06_01_biz.pdf.

Korinek, J. and P. Sourdin (2009), "Clarifying Trade Costs: Maritime Transport and its Effect on Agricultural Trade", *OECD Trade Policy Papers*, No. 92, OECD Publishing, Paris, http://dx.doi.org/10.1787/220157847513-en.

Maritime Insights (2014), "Container throughput and economic development", *Container Throughput in Ports in the Philippines, Singapore, Thailand and Viet Nam*, Vol. 2/3, autumn, https://www.polyu.edu.hk/lms/ICMS/research_maritimeInsight/2014-Autumn-en/1.pdf.

Merk, O. et al. (2011), "Competitiveness of port-cities: The case of the Seine Axis (Le Havre, Rouen, Paris, Caen) – France", *OECD Regional Development Working Papers*, http://dx.doi.org/10.1787/5kg58xppgc0n-en.

Ministry of Transport (2011), *Viet Nam Seaport Master Plan 2020 - With Orientations Toward 2030,* Ha Noi, Viet Nam.

Miola A. et al. (2009), *External Costs of Transportation. Case Study: Maritime Transport. European Commission*, Joint Research Centre, Institute for Environment and Sustainability, pp. 1-109, http://publications.jrc.ec.europa.eu/repository/handle/JRC51275.

Mongelluzzo, B. (2012), "How California's ports cleared the air", *Journal of Commerce*, 23 January, www.joc.com/portsterminals/how-californias-ports-cleared-air?page=2, (accessed 23 July 2016).

Nordas, H., E. Pinali and M. Geloso Grosso (2006), "Logistics and Time as a Trade Barrier", *OECD Trade Policy Working Papers*, No. 35, OECD Publishing, Paris, http://www.oecd.org/officialdocuments/publicdisplaydocumentpdf/?doclanguage=en&cote=td/tc/wp(2006)3/final.

Notteboom, T. and J.-P. Rodrigue (2011), "The corporate geography of global terminal operators", in: J. Kujawa and O. Debicka (eds.), *Development and Functioning of Enterprises in Global and Changing Environment*, The Foundation of the Development of Gdansk University: Gdansk, pp. 160-180.

O'Brien, T. (2004), "Quality of life and port operations: Challenges, successes and the future", white paper prepared for the sixth annual CITT State of the Trade and Transportation Industry Town Hall Meeting, Center for International Trade and Transportation, Long Beach, California.

OECD (2014), *The Competitiveness of Global Port-Cities*, OECD Publishing, Paris, http://dx.doi.org/10.1787/9789264205277-en.

OECD (2011), "Green Growth Potential of Port Cities: Improving Your Green Port Positioning", OECD workshop presented by E. Haezendonck, 13 June, Paris.

OECD/ITF (2015a), "The Impact of Mega-Ships", *International Transport Forum Policy Papers*, No. 10, OECD Publishing, Paris, http://dx.doi.org/10.1787/5jlwvzcm3j9v-en.

OECD/ITF (2015b), *Port Investment and Container Shipping Markets*, OECD Publishing, Paris, http://dx.doi.org/10.1787/9789282107850-en.

Thanh Nien News (2010), "Industrial toxins found in 15 containers at Hai Phong port", www.thanhniennews.com/society/industrial-toxins-found-in-15-containers-at-hai-phong-port-15781.html (accessed 31 May 2016).

Tran, D. L. et al. (2013), "Assessing environmental conflicts in Viet Nam: Case studies of Hai Phong and Nha Trang City", Chapter 9, in *Environmental Conflicts in Coastal Urban Areas: Towards a Strategic Assessment Framework for Sustainable Development,* Sapienza Università Editrice, Rome.

USAID (2014), "Sister Cities Disaster Preparedness Program", https://www.usaid.gov/sites/default/files/documents/1861/FS_SisterCities_Eng.pdf USAID Viet Nam.

Viet Nam News (2015), "Banned industrial waste leaking in via import loopholes", http://vietnamnews.vn/environment/265442/banned-industrial-waste-leaking-in-via-import-loopholes.html (accessed 31 May 2016).

Viet Nam Seaports Association (n.d.), www.vpa.org.vn/english/information/info_static2014.html (accessed 10 November 2016).

VIR (2016), "Hai Phong to Capitalise on a Good Year", Viet Nam Investment Review, Ho Chi Minh City, Viet Nam, www.vir.com.vn/HaiPhong-to-capitalise-on-a-good-year.html (accessed 30 June 2016).

VIR (2013), "Vinalines drops plan for PPP port investment", Viet Nam Investment Review, Ho Chi Minh City, Viet Nam, http://www.vir.com.vn/haiphong-to-capitalise-on-a-good-year.html (accessed 18 June 2016).

Wilmsmeier, G. and R. Sanchez (2009), "The Relevance of International Transport Costs on Food Prices: Endogenous and Exogenous Effects", *Research in Transport Economics*, Vol. 5(2), pp. 56-66.

World Bank (2014), "Container port traffic", *World Development Indicators*, http://data.worldbank.org/indicator/IS.SHP.GOOD.TU/countries (accessed 20 July 2016).

Chapter 4

Adjusting urban governance for green growth in Hai Phong

Chapter 4 examines urban governance for green growth in Hai Phong and is structured into the following six sections:

1) Introduction: The governance context in Viet Nam

2) Co-ordination mechanisms between government bodies

3) Financing mechanisms

4) Administrative capacity

5) Mechanisms to engage citizens and civil society

6) International co-operation.

Main points

- **Hai Phong's green growth will depend to a large extent on the effectiveness of its governance structure and the way government operates.** In Viet Nam, the central government has been actively promoting green growth, recognising the role cities can play. The *Doi Moi* economic reforms have transformed the economic landscape of Viet Nam, which could also work positively for green growth in cities. However, transforming Hai Phong into a "green port city" requires a higher calibre of urban governance, with a clear division of responsibilities and dynamic co-ordination across levels of government.
- The national government has a key role in providing suitable **legal and policy frameworks** and support for investments to ensure that Hai Phong achieves its green growth goals. National strategies and policies should be reviewed and revised to encourage greater recognition of interdependence between central, provincial and local governments, and among Hai Phong and neighbouring cities. The channels of communication across levels of government need to be kept open and active.
- Hai Phong's green growth plans are linked to those of other cities and jurisdictions and national activities. Co-ordination for service delivery, building of infrastructure and proposals for common goals for urban development could greatly enhance green growth policy performance. For better co-ordination among its peers, Hai Phong authorities, and the central government, may wish to invest in a **more collaborative approach to governance that underpins regional/metropolitan planning.**
- Green growth requires **cross-sectoral planning** to tackle issues comprehensively. Close co-ordination between the Port Authority and Hai Phong's city government is also critical for achieving green growth targets and better public investment. For instance, port activities are not covered in depth in the Hai Phong Green Growth Promotion Plan, despite the vision of this strategic document to transform the city into a "green port city". Integrated territorial approaches for development in a city have a more positive impact than segmented policies; for instance, infrastructure investment works better when associated with education and skills, and land-use policies with transport and environment policies.
- Hai Phong's green growth action plan appears to be constrained by a lack of financial resources. To build solid ground for financing green growth, Hai Phong should rely more on **own-source revenues**, including taxes and fees to pay for services, with the support of the national government. The hierarchical fiscal structure of Viet Nam limits the possibility for joint projects between cities in the same region.
- To enhance the capacity of the local public administration, Hai Phong needs to invest in the **skills and competences of the public workforce**, encouraging high ethical standards to prevent corruption in critical areas such as the expansion of the seaport, and improving regulation and dealing with red tape to facilitate investment in green growth projects.
- Achieving green growth objectives requires **citizen engagement in decision making**. Hai Phong needs to strengthen its relations with citizens by providing access to reliable and timely information and promoting a wider participation of communities in the planning and implementation of green growth strategies. Ensuring that the **legal and regulatory framework** considers consultation with citizens as part of the decision-making process and **developing guidelines** on how to engage with civil society would be critical. This would also help raise public awareness and understanding of the benefits of urban green growth.
- Hai Phong has benefited from almost USD 70 billion in **official development finance** between 2002 and 2014, the second-largest recipient in Southeast Asia, after Indonesia. It has also benefited from such financial support for port development and water supply and sanitation systems (around USD 500 million in the same period). The effectiveness of this aid could be enhanced, however. Official development finance for the port should, as far as possible, support a comprehensive package including the port's hinterland connections and its environmental performance.

4.1 Introduction: The governance context in Viet Nam

National policy framework for green growth

In Viet Nam, the central government has been actively engaged in green growth. In 2012, the Vietnamese government produced the **National Green Growth Strategy** for the period 2011-20, with a vision to 2050, and in March 2014 approved the **National Action Plan on Green Growth** for the period 2014-20 (Box 4.1). Notably, the strategy is legally binding and was approved by the prime minister, after being developed by an inter-ministerial working group led by the Ministry of Planning and Investment, which is also in charge of the national development plan. Moreover, the National Action Plan assigns the Ministry of Finance to develop a framework to mobilise finance for the strategy equally from the national budget and international sources. The strategy is supported by high-level leadership, full country ownership and the engagement of national ministries (OECD, 2014).

The central government also recognises the roles that cities can play in encouraging green growth. Among the 12 measures laid out by the National Action Plan, Action 11 (to develop green and sustainable urban areas) specifically targets urban actions. In addition, Action 2 requests some provinces and cities to formulate a local Green Growth Action Plan, implement small-scale green pilot projects and to synthesise and disseminate best practices, with the Ministry of Planning and Investment as a co-ordinating agency.

In 2014, based on these national initiatives, Hai Phong adopted the **Green Growth Strategy Action Plan**. This aims to make Hai Phong a **"green port city"**, taking advantage of the historical port city and setting up a modern, competitive and eco-friendly green port system. In 2015, the city also produced the **Green Growth Promotion Plan**, with the support of City of Kitakyushu, Japan, to identify tangible projects to implement the Action Plan.

These national and local initiatives are an indication of Hai Phong's promising future for green growth. However, **achieving the ambitious goal of "green port city" will depend, to a large extent, on the effectiveness of its governance structure and the way government operates.** The experience of OECD countries suggests that a key criterion for effective implementation of green growth strategies is how well these are supported by the dominant, political, legal and bureaucratic structures and processes. This requires making sure that decisions and programmes implemented in one area are aligned or at least not in conflict with programmes in other areas. It is worth noting that the responsibilities and capacities of Hai Phong, shaped by the other levels of government, are frequently over-stretched.

Box 4.1. **Viet Nam Green Growth Strategy (VGGS) and National Action Plan**

The Viet Nam Green Growth Strategy (VGGS) aims to achieve sustainable economic development by reducing carbon emissions, increasing ability to absorb greenhouse gases (GHG) and enriching natural capital. The strategy outlines short, medium and long-term objectives, prioritised for implementation, as well as a vision for 2050. It is anchored in three strategic tasks:

1. **Low-carbon growth**: the VGGS sets out systematic targets starting from the period 2011-20 through 2030 to 2050 in hopes of reducing the intensity of GHG emissions per unit of GDP (8-10% or double the target with international support) and encourages the use of clean and renewable energy.
2. **Greening of production**: a "clean industrialisation'' strategy will green existing business and create new ones, adopting environmentally friendly technology, preventing and remediating pollution and investing in natural capital.
3. **Greening of lifestyle**: the traditional lifestyle is to be combined with modernised yet comfortable, high-quality living standards that are rooted in nature, as experienced in the rural areas, and at the same time in tune within the context of global integration. The focus is on promoting green growth, expanding public transport and effective waste collection and treatment.

Ten priority programmes and projects were outlined for implementation from 2011-15.

After the VGGS, the Prime Minister's Office launched the **National Green Growth Action Plan** for 2014-20. The National Action Plan designates 12 actions: i) formulating institutions to support the implementation of VGGS; ii) localities formulate local Green Growth Action Plan and implementing small-scale green pilot projects; iii) implementing efficient use of energy and reducing GHG emissions in some energy-consuming industrial sectors; iv) implementing efficient and effective use of energy and reducing GHG emissions in the transport sector; v) changing cultivation techniques to reduce GHG emission in agriculture, forestry and aquaculture; vi) developing clean and renewable sources of energy; vii) reviewing development strategies, master plans and plans and formulating a programme for restructuring the economy towards for green growth; viii) efficient and sustainable use of natural resources and developing green economic sectors; ix) developing sustainable infrastructure; x) promoting the "Enterprises for Sustainable Development" campaign and improving capacity, as well as developing technical and management services market for green growth; xi) developing green and sustainable urban areas (cities); and xii) promoting a green lifestyle.

Each action is supplemented by 65 activities, of which 23 are listed as priorities. One such activity is an inter-ministerial **Co-ordinating Board for Green Growth Strategy**, which was established under the National Committee on Climate Change to implement the strategy. Another priority is to formulate **Local Green Growth Action Plans** in some provinces and cities.

Source: Government of Viet Nam (2012a), *Viet Nam National Green Growth Strategy for the Period 2011-2020, with a Vision to 2050* https://www.giz.de/en/downloads/VietNam-GreenGrowth-Strategy.pdf; Government of Viet Nam (2014), *National Action Plan on Green Growth for Period 2014-2020.*

Doi Moi economic reforms provide opportunities for Hai Phong

The *Doi Moi* economic reforms have transformed Viet Nam's economic landscape, as well as city administration, which can have a positive impact on green growth. According to the Viet Nam Trade Promotion Agency (2014), the reform has allowed Viet Nam to sustain a growth rate of around 7% between 1990 and 2012. [1] The *Doi Moi* reforms, as translated into city administration, mean that the city has greater autonomy to meet citizens' needs for services and infrastructure, and collecting taxes and fees to finance them, to carry out these functions in an open and transparent way, and then to be held accountable for the results. The reforms have laid the foundation for a deeper transformation of the country's economic objectives, and have the potential to underpin the transformation of Hai Phong (Box 4.2).

Despite their positive impact in Hai Phong, the *Doi Moi* reforms have also led to patchy or asymmetrical decentralisation in the main provincial cities, which suggests some areas for further improvement. While Ho Chi Minh City abides by the regulations issued by central government, cities such as Da Nang "break barriers" to mobilise resources for development (Nguyen et al., 2013). Hai Phong appears to be in a similar situation. Cities under central government have no more authority in fiscal policy than rural local governments. People's Committees of districts and wards have autonomy in allocating the distributed budget. Ho Chi Minh City receives an award if the revenues to the national budget exceed target estimates. The city is awarded 30% of the amount exceeded, but this amount may not be greater than any reward issued in the previous year. Hai Phong is not part of that reward mechanism.

Ultimately, shifting to a green port city is likely to require a higher calibre of urban governance than the pursuit of green growth alone. Achieving Hai Phong's potential advantages for environmental improvement, raising living standards and social equity will depend on adequate urban governance structures with a clear division of responsibilities and strong and dynamic co-ordination and co-operation mechanisms with other cities and across levels of government.

Box 4.2. **Viet Nam's *Doi Moi* economic reform**

In 1986, the Vietnamese government introduced the *Doi Moi* (renovation) economic reform. Its aim was to direct effort to the development of agriculture, the expansion of the production of consumer goods, and the promotion of trade and foreign investment relations. *Doi Moi* abolished the system of bureaucratic centralised management based on state subsidies, and moved to a multi-sector, market-oriented economy with room for the private sector to compete with the state in non-strategic sectors. This was a signal that the country was moving away from central planning towards indicative planning and macroeconomic policy levers. A key change was the new emphasis placed on the role of government in the industrialisation process. The state was to concentrate on building the necessary premises for the acceleration of socialist industrialisation in the subsequent stage. As part of the reforms, a Law on Land was passed in 1987. Its main premise was that while the state retained ownership of land, private land-use rights were recognised by the state. Viet Nam's reform process was accompanied by a period of liberalisation and international integration, through which the market economy was developed.

The top-down approach to planning in general changed fundamentally with *Doi Moi*. In 1989, decisions on the socio-economic development plans of the local levels were decentralised, and in 2003, local governments of all levels gained more responsibilities in the co-ordination, budget allocation and finalisation of the socioeconomic development plans in their respective localities. The central authorities, the Ministry of Planning and Investment and the Ministry of Finance, now only provide the general development orientation and the budget constraints, and planning agents (provinces and line ministries) prepare their detailed plans and submit them to the Ministry of Planning and Investment. In that sense, planning is less top-down than in the past.

Source: van Arkadie, B. and R. Mallon (2003), *Viet Nam: A Transition Tiger? The Introduction of Doi Moi*; World Bank (2010), *Viet Nam Development Report 2010 – Modern Institutions*, www-wds.worldbank.org/external/default/WDSContentServer/WDSP/IB/2010/03/19/000333037_20100319004055/Rendered/PDF/535730WP0P11221ng0reduced0file0size.pdf.

4.2. Co-ordination of green growth between government bodies

Governance reforms to improve relationships across levels of government and co-ordination and co-operation among cities are at the core of achieving green growth objectives. Viet Nam's national government and Hai Phong will need to revise the current governance arrangements to ensure they are facilitators of transforming the city into a "green port city". For Hai Phong, what matters most is the quality of the co-operation mechanisms across and among levels of government, given the high dependency on the central government. The specific profile of each port-city also requires a high degree of collaboration between private port authorities and local government, to integrate port activities into local green growth strategies.

Improving relationships across levels of government

As in OECD countries, the national government of Viet Nam has a critical role to play in ensuring that Hai Phong achieves its green growth objectives. Key functions of the national government include providing suitable legal and policy frameworks, supporting investments in appropriate forms of green infrastructure, and helping to ensure that Hai Phong's shift towards green growth can succeed politically, economically, socially and environmentally. The national government can pass legislation and offer fiscal incentives to promote regional cohesion and joint work or investment in, for example, green urban infrastructure. Such instruments are best devised and implemented nationally. The National Green Growth Action Plan proposes as priorities reviewing and revising national sectoral strategies and policies (e.g. energy) to pursue green growth. At the same time, such **national strategies and policies should encourage greater recognition of interdependence between central, provincial and local governments, and among Hai Phong and neighbouring cities** on the role they have in pursuing green growth objectives.
The key message is that green growth is more likely to be achieved through partnerships among governments and not only through processes involving government and non-government stakeholders.

To nurture collaboration between Hai Phong and the national government, **the channels of communication between spheres of government need to be kept open and active**, to facilitate the exchange of information. According to the World Bank's 2010 Viet Nam Development Report, little communication exists between investment budget planners, the Ministry of Planning and Investment, provincial departments of planning and investment and those that plan the current budgets, the Ministry of Finance and provincial departments of finance. This means that there is little linkage between recurrent and capital budgets. At present, local planning generates a wish list, rather than a strategic vision for management. Sharing or unclear lines of authority makes some decision-making processes more cumbersome than they need to be. This has negative implications for urban development and green growth initiatives.
For instance, the Department of National Resources and Environment (DONRE) issues land use right certificates, ownership of houses and other assets attached to the land. However, the issuance of land use certificates has often been an area of concern due to the slowness of the process. Similarly, transparency in land use plans and pricing has also often not been clear which is collaborated by Hai Phong's PAPI ranking of 44th out of 63 provinces under the transparency measurement. The national government could also clarify the responsibilities of the different levels of government in greater detail. Viet Nam's 2002 State Budget Law and the 1996 Ordinance on the Specific Tasks and

Powers of the People's Councils and People's Committees at each level try to clarify roles and authority. However, in many cases, central government decisions are not elaborated as guidelines and therefore cannot be implemented. For green growth, it is not yet clear how local responsibilities are defined, although both the NGGS and the National Action Plan appear to recognise the role of cities and provinces in achieving objectives set nationally. To ensure sub-national green growth actions, the national government could consider enacting **a legal framework**, along with laws/regulations of People's Committees and People's Councils, to provide the basis for political, administrative and fiscal relations for each level of administration. This would give them clear mandates for green growth in different sectors. Legislating on better inter-governmental relations to continue the decentralisation process initiated with the *Doi Moi* could also clarify the specific expenditure responsibilities of provincial/lower levels of government, their authority to levy taxes and fees to finance green investment, and the extent to which provinces and local governments can borrow or increase debt. Hai Phong would benefit enormously from this, because it would have more leverage to decide on expenditure and investment priorities for implementing green growth projects. In addition, a **National Strategic Reference Framework (NSRF)** could provide insight on a basis for dialogue and co-ordination on green growth initiatives at the national level. Such a framework could introduce ways to further co-ordinate and guide local green growth initiatives. In Sweden, for example, the design and implementation of an NSRF has led to improved co-ordination among levels of government with private actors and across sectors on different policy areas: competitiveness, entrepreneurship and employment (OECD, 2010b). A Vietnamese version of NSRF could establish thematic groups to promote green growth.

Enhancing co-ordination among neighbouring provinces through metropolitan planning

Hai Phong's green growth initiatives are linked to other cities/jurisdictions and national activities. A key criterion for the effectiveness of green growth strategies is how well these are co-ordinated with neighbouring cities and lower levels of government under a metropolitan vision. Little evidence suggests that Hai Phong's Local Green Growth Action Plan is based on comprehensive metropolitan planning beyond the jurisdiction of the City of Hai Phong. Its focus is within the city, without any indication of how the city will implement the plan in relation to other neighbouring communities. Urban development in Hai Phong has had long-term environmental consequences that are often cross-border. Such spillovers, in general, make independent planning and the Green Growth Action Plan somewhat inefficient.

The features of the local administrative system in Viet Nam suggest that it is a highly hierarchical system, with lower levels of government co-ordinated through the central government (Figure 4.1). Hai Phong is one of the five provincial cities of Viet Nam directly supervised by national central government. The other four cities are: Ho Chi Minh City, Ha Noi, Da Nang and Can Tho. These traditional arrangements are complex, and may lead to duplication of responsibilities and setting contradictory goals for the same areas at the moment of planning. The direct line of guidance from each ministry to each city department has also confused the decision-making process, and reduces the incentive to take the initiative. In addition, the devolution of authority to lower levels of government as a result of the *Doi Moi* reforms has resulted in competition between provinces, including centrally run cities. Competition can encourage a race to the top, especially in creating a favourable business environment and making progress on green growth. But where the decisions of one province affect another, whether positive or negative, lack of regional co-operation can lead to inefficiencies.

Figure 4.1. **Viet Nam's regional and urban administration hierarchy**

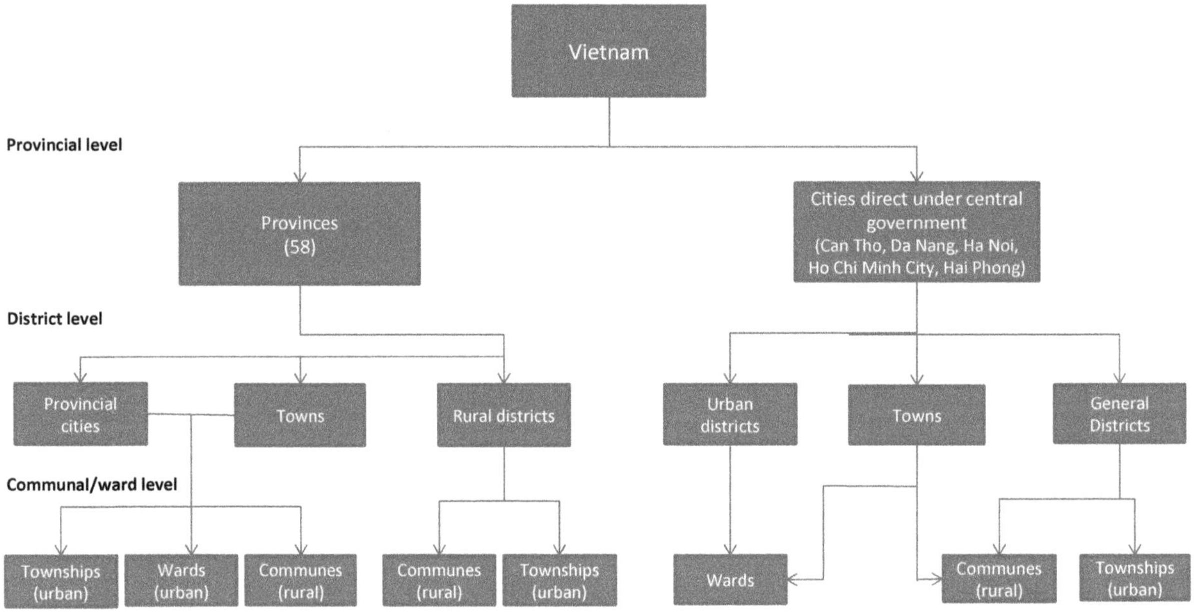

Note: Hai Phong has 7 urban districts, 6 rural districts and 2 island districts (Cat Hai, Bach Long Vi).

Source: Calculations based on World Bank (2011), *Viet Nam Urbanization Review*, https://openknowledge.worldbank.org/bitstream/handle/10986/2826/669160ESW0P1130Review000Full0report.pdf.

To achieve Hai Phong's green growth objectives, strengthening relations among local governments and other actors is essential in a context of limited resources. Co-ordination for service delivery, building of infrastructure, or even proposals for common goals for urban development could substantially enhance green growth policy performance. To encourage green growth through better co-ordination among its peers, Hai Phong authorities, together with the central government, may wish to invest in a **more collaborative approach to governance that underpins regional/metropolitan planning**. The following options could be considered:

- The national government, with the support of provinces and province-level cities, could develop the concept of **functional urban areas (FUAs)** in Viet Nam and a metropolitan planning framework to build synergies for green growth. The Ministry of Planning and Investment and relevant national ministries can help build mechanisms for collaboration across levels of government to improve the strategic metropolitan planning framework. The experience of OECD countries (e.g. Canada, France, Germany, and Japan) suggests that there is no optimal degree of decentralisation. What matters more is the quality of co-operation mechanisms across levels of government, given the high interdependency among the different levels. OECD countries have realised that metropolitan planning is critical for cities to grow sustainably. If Hai Phong and its surrounding communities are to become functional as a region, they need to plan growth as an FUA. Metropolitan planning would permit them to form a coalition to achieve objectives that have been collectively agreed upon by applying their joint authority. It would also help Hai Phong to connect the values and needs of the population with the city's policy goals. The OECD's methodology could be applied to define FUAs in Viet Nam. In this vein, Hai Phong needs to create new processes for formulating competitiveness strategies at the functional scale of the territory. In this respect, studying the example of France's spatial planning could be useful. In France, sub-national governments have formulated the Territorial Coherence Scheme (*Schéma de cohérence territorial*, SCoT), to ensure a balance between urban renewal and the rural periphery, to preserve the diversity of urban functions and social mix, and to achieve competitive and sustainable development (Box 4.3).

- **Consider metropolitan planning and the creation of a metropolitan planning institute.** Land-use, transport, service delivery and green growth investment initiatives are best addressed taking the city core and suburban area as a whole. Hai Phong and its neighbouring cities could engage together in metropolitan/regional planning. This could be the basis for establishing co-operative networks where information can flow through a joint action framework. Effective metropolitan planning, to bring lasting change, must involve a large number of stakeholders and be backed by long-term finance. This does not necessarily imply a uniform approach, as the challenges Hai Phong faces may differ from those of other neighbouring cities. Hai Phong could lead metropolitan planning by creating a metropolitan planning institute, which could act as the regional planning authority. Its purpose should be to improve the effectiveness of economic development and green growth by co-ordinating planning and implementation. The planning institute could prepare long-term plans, provide technical assistance, propose integrated metropolitan development projects and prepare mechanisms for evaluation. The planning institutes of Barcelona (Spain) and Guadalajara (Mexico) could prove an inspiration here (OECD, 2015a).
- Green growth could have positive spillovers in the region if proper planning and infrastructure are in place. Infrastructure planning in Hai Phong is one clear instance where regional planning can improve outcomes, due chiefly to the seaport expansion and modernisation and the railway currently under construction. Interviews for this case study indicated that these two major infrastructure projects were launched independently without close study of the developments in the region or the plans and projects adopted by other ministries. This risks duplication of efforts. Building synergies with other municipalities, cities or provinces to pursue green growth is critical, since Hai Phong alone does not have the critical mass of population or economic activity to sustain the port. But if neighbouring coastal provinces, such as Thai Binh, Quang Minh and Nam Dinh and even Hai Duong, can co-operate, they could sustain and benefit from Hai Phong's port. Similarly, planning for the roads and railway to Ha Noi can increase the opportunity for both to benefit. If the northern coastal provinces can agree where Hai Phong's port should go and how the road and rail networks should be organised, the chances of achieving green growth will increase. In this respect, the national government can facilitate metropolitan planning, even if it does not direct it. **Providing funds for multi-province initiatives** is one solution, which underpins such schemes as the metropolitan fund in Mexico.[2]
- **A high-level forum for co-ordination and collaboration** could be established at which senior members of government from the different cities can meet, discuss issues and agree on joint approaches and projects in open and consensus-based processes. A senior executive programme for public and private senior managers from across the metropolitan area could be established so they could be trained,

discuss the underlying issues and agree on a shared understanding and vision for green growth in the region. Seminars, discussion forums, learning cafés and workshops are other options for promoting collaboration and network building. Building a metropolitan network of senior officials on green growth and metropolitan planning is key; such issues cannot be left to lower levels of the hierarchy, which have a narrower focus. One common element in the success of service delivery integration in such countries as Australia, Canada, Belgium, Denmark and the United Kingdom has been the active engagement of political leaders.[3] Greater political engagement allows the formation of the collaborative political mechanisms needed. Building shared, seamless governance models can help to align policies across levels of government.

Box 4.3. **The Territorial Coherence Scheme of France (SCoT)**

The Territorial Coherence Scheme (SCoT) is a tool for conception and implementation for inter-communal strategic planning for wide catchment areas or urban areas, within the framework of a development project and sustainable development. City plans (*plans local d'urbanisme*), local urban transport plans and housing plans must be compatible with the SCoT to be valid and enforced. The SCoT sets the main orientations of the organisation of a group of adjacent communities (*intercommunalité*) for a 10-year period. The scheme must respect principles of sustainable development: the principle of equilibrium between urban renewal, urban development, rural development, and the preservation of natural spaces and landscapes; and the diversity of urban functions and social mix; and the principle of respect for the environment.

Building a SCoT is decided at the initiative of the municipality or a coalition of municipalities. The decision is made by a resolution of each municipal assembly in the area, with a two-thirds majority. The coalition of municipalities is the main actor, and the national government, the region and the prefecture can participate. In addition, opinions are heard from the prefect, the chair of the Regional Assembly, the chair of the Departmental Assembly and the mayors of the municipalities. After taking the necessary steps in the public hearing system, the SCoT is approved by the coalition of municipalities. The SCoT consists of i) a diagnostic and environmental assessment; ii) a project of development and sustainable development (*projet d'aménagement et de développement durables*, PADD); and iii) the document with orientations and objectives.

By January 2015, 448 SCoTs had been approved or were in the process of being approved, covering 25 137 communes (almost 70% of the communes in the country), 50.5 million inhabitants (77% of the national population), and approximately 60% of the national territory. After the passage of the Law for the National Commitment to the Environment (*loi d'Engagement National pour l'Environnement*, or ENE) in 2010, the SCoT was reinforced, giving priority to efficient management of the space; enlargement of the scope of the SCoT to new areas (i.e. development of electronic communications, reduction of GHG emissions, and preservation and restoration of ecological continuity); densification; and development of regional schemes of ecological coherence.

Source: Government of France (2015), *Schéma de Coherence Territoriale* (SCoT), *Ministère du Logement, de l'Égalité des Territoires et de la Ruralité*, http://www.logement.gouv.fr/schema-de-coherence-territoriale-scot (accessed 3 August 2016).

Improving co-ordination across agencies within Hai Phong's government

The interviews conducted for this case study suggest that in Hai Phong, local departments, agencies and other administrative units tend to work in their comfort zones, rather than in a cross-sectoral setting, which leads to duplication and limited joint efforts for addressing common issues. Line departments, as in other levels of government, display a strong command-and-control and interventionist orientation, in part based on interests in controlling revenues through the state-owned enterprises attached to them (Fritzen, 2006).

Promoting green growth requires **cross-sectoral planning** to tackle issues in a comprehensive manner. One of Hai Phong's key urban transport development projects is the construction of a trunk road on the east-west transport corridor from Bac Son commune to Nam Hai ward, connecting Hai Phong's seaport to the motorway. The aim is to improve traffic for vehicles travelling in the city and transporting goods from outside to Hai Phong port system.[4] It is not clear how it will contribute to green growth and other environmental objectives. Without clear environmentally sound transport strategies in addition to the road strategy, it sends the message that Hai Phong is focusing more attention on traditional vehicle-oriented transport policies than on more sustainable forms such as railways.

Close co-ordination between the **Hai Phong Port Authority** and city government is also critical for achieving green growth targets and better public investment, given the huge impact it will have in the city and region. This must allow for the incorporation of the port renewal and expansion plans into the urban development plans and environmental strategies. A process of urban strategic planning aligned with the expansion project of the port can lead to greater consistency in policy action on the part of local authorities. As explained in Chapter 3, the port's specific green growth challenges are not a central focus of the Hai Phong Green Growth Promotion Plan and Action Plan, even though the "green port city" is the central vision of both strategic documents. The green port recommendations detailed in Chapter 3 should therefore be integrated into these documents, and Hai Phong Port Authority and Hai Phong city government must work together to ensure implementation. Given the strong involvement of the national government in port development, it would also be useful to integrate green growth into the Viet Nam Seaport System Master Plan. For a longer-term approach, the port expansion plans should develop a regional perspective co-ordinated across relevant agencies.

Co-ordination should start in the early stages, as the construction of a new terminal would present logistic challenges. For example, access is necessary at the construction phase of the new terminal to allow the materials and equipment to get to the construction site. Building the new terminal will require services (i.e. technological, materials and human capital) that can only be achieved through co-ordination. The new seaport terminal is likely to result in a new centre of economic and social activity, which will present both risks and challenges. Co-ordination between the port authority and Hai Phong's city government will be needed to ensure adequate services, such as schools, hospitals, housing, recreation centres, transport, waste management and public spaces. These scenarios need to be considered as a part of urban planning and green growth strategies. Local authorities should also take this opportunity to inform residents in the seaport expansion area of its main positive and negative impacts.

Lack of skills and incentives, and of any formal mechanism for working together, is a major challenge to establishing and strengthening horizontal linkages across agencies. In the case of Viet Nam, the 2009 Law on Urban Planning does not specify any procedure for cross-sectoral co-ordination of urban planning at the city level. The authority responsible for urban planning is requested to collect comments from relevant stakeholders, but no clear incentive or mandate is in place to engage in integrated urban planning. Moreover, holding a single annual planning meeting to discuss the Socio-Economic Development Plan (SEDP) at the provincial level may not be enough to reach consensus on how to encourage socio-economic development with a cross-sectoral approach. Another constraint for establishing inter-sectoral linkages is the capacity of staff with the skills to conduct cross-sectoral planning. Line ministries tend to be predisposed towards controlling and/or monitoring inputs, rather than processes and outcomes. They also tend to have difficulty adapting to the new roles they must play in a decentralised system, such as enforcing minimum standards and providing technical support to lower levels of government (districts and communes) (Fritzen, 2006). Demarcation of mandates of various agencies leaves room for discretion in the interpretation of responsibility and accountability, where overlapping functions and duties prevail. Interviews for this case study showed that centralised control by line ministries over policy reduce the incentive for department heads to innovate, achieve greater efficiency and improve productivity, which can limit green growth.

The experience of OECD countries suggests that integrated territorial approaches to urban development have a more positive impact than segmented policies. For instance, infrastructure investment works better in association with education and skills, and land-use policies with transport and environment policies. The following practical options could be considered:

- ***Establish a Green Growth Co-ordination Council***, including a decision-making hierarchy and partnership for achieving better results in the implementation of the Hai Phong Green Growth Local Action Plan. The Co-ordination Council could be inspired by the Viet Nam Urban Forum at national level, which gathers representatives of the public and private sector as well as international organisations to discuss urban development challenges (Box 4.8). The difference would be that the Co-ordination Council would have decision-making power and a forum to define policy at provincial-regional level. These co-ordinating arrangements and structures could be used as platforms for agreeing upon and reviewing green growth policies and programmes; mobilising resources; co-ordinating multi-sectoral partnerships and consulting with a wider group of stakeholders. The structure of the Co-ordination Council would have three main levels: executive, technical co-ordination and thematic working groups, along with a consultative advisory group (Figure 4.2).

  The council would be headed by an executive committee to provide general oversight. Central and local policy leaders will constitute the executive committee. The co-ordinating council would act as a secretariat and constitute a network of department heads, from, for example, the departments of Planning and Investment, Natural Resources and Environment, Finance, Transport and Construction. It would also include representation from the seaport, because the strategy for greening the port's activities should be an essential part of the green growth. Civil society organisations and private sector representatives should also be brought on board to pool practical knowledge, identify and anticipate policy issues and practical problems and come together to develop a consensus position. The involvement of the central government in the co-ordination council would be desirable, because it has the authority to grant or divert resources and arrange co-ordination. Under the co-ordinating council, technical working groups could be established in areas of strategic importance to the green growth vision, such as the port, land use and transport, etc. The working groups would include experts and stakeholders, who would consider public participation and inputs in the production of its deliverables. Another option for Hai Phong authorities is to ***form a joint reviewing board for green growth*** composed of the finance, planning and investment, construction and environment departments, to review all green growth projects and ensure that they are aligned with projects with other departments. Such a board can also ensure projects are financially viable. The People's Committee could oversee the functions and performance of the Green Growth Co-ordination Council or the joint board to ensure that all decisions are within the framework of the law.

- ***Make a clear and sound legal framework to ensure co-ordinated action***. No one solution or right formula is available to harmonise competing interests, overcome uncertainty and ambiguous government structures and make hard policy choices that everyone will accept (Laegrid, Dyrnes Nordo and Rykkja, 2013). The greatest likelihood for cross-sectoral co-ordination seems to occur when legal mechanisms are implemented to force policy sectors to address the interests of another sector (Dudek, 2005). No sector is likely to be successful without co-ordinated action with others. The experience of OECD countries suggests that for horizontal co-ordination to achieve results, it must be carried out across the different levels of government by government itself. In that sense, Hai Phong could consider formally allocate the leadership and decision-making role to the co-ordinating council in the legislation. The council has a key role to play in ensuring vertical and horizontal co-ordination, and legislation could encourage delivery of concrete results, establish accountability and work to achieve shared objectives for green growth.

Figure 4.2. **Proposal for a Green Growth Co-ordination Council for Hai Phong**

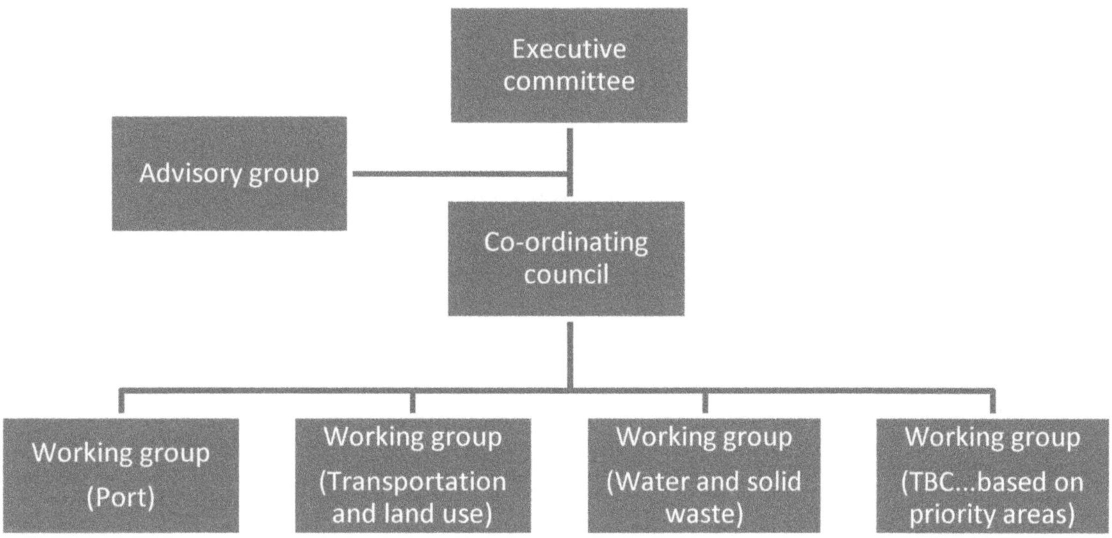

Source: Own elaboration.

4.3. Financing green growth in Hai Phong

Ensuring sound financial ground for paying for green growth and urbanisation

The pressure on Hai Phong to spend on green growth initiatives is not likely to lighten up in the near future. The demand for public services will continue to rise as the city grows. It is thus of critical importance for Hai Phong to ensure sustainable sources of funding to pay for green growth initiatives and urbanisation.

The budget of the city of Hai Phong has dramatically increased, by more than five-fold over the past decade, from VND 8 462 billion in 2004 to VND 50 100 billion in 2014. The total real budget more than doubled, from VND 8 462 billion in 2004 to VND 19 293 billion, over the same period. Most of the revenue has been generated through customs revenue (73.1% of total revenue in 2014) received from port activities, while domestic revenues (i.e. agricultural land-use tax) only account for a minor share (19% in 2014) (Figure 4.3). The increase in total revenue is explained by the increase in custom revenue, in line with the dramatic growth in port activities in the past 15 years. Customs revenue, however, dropped in 2012, due to a sudden slowdown of seaport activities in the country that drove total revenue down temporarily. Other sources of revenue include mainly transfers from the central government (6.8% in 2013) (Table 4.1). This is very low compared to many other cities in Southeast Asia: transfers from the central government of Indonesia, for instance, accounted for 44% of Bandung's local budget in 2016 (OECD, 2016, forthcoming). This can be explained by the relatively high budget revenue of Hai Phong, thanks to customs revenue: transfer revenues in Viet Nam are partly based on the capacity of local governments to generate their own revenue – which is considered high in the case of Hai Phong.

Despite a steady increase in the real budget in recent years, the financial capacities of the local government of Hai Phong are limited. Hai Phong only received USD 334 per inhabitant in 2014. In comparison, Bandung and Bangkok received USD 652 and USD 1 075, respectively (Figure 4.4). This gap in financial resources may undermine the city's efforts to achieve green growth, and should be bridged.

Table 4.1. **Breakdown of Hai Phong's budget (2013)**

Type of revenue	Item	Value (VND billion)	Share of total revenue (%)
Domestic revenue	Total domestic revenue	8 540.149	19.00
	Agricultural land-use tax	5 889.378	13.10
	Personal income tax	0.146	0.00
	Registration fee	725.077	1.60
	Fees and charges	207.802	0.50
	Land revenue	704.124	1.60
	Other revenue	978.162	2.20
Customs revenue	Total customs revenue	32 876.046	73.10
	Import tax	10 466.585	23.30
	VAT on imports	22 409.461	49.80
Other	Total other	3 606.758	8.00
	Aid revenue (excluding aid for lending)	338.044	0.80
	Revenue from budget balance of previous year	17.262	0.00
	Transfers from national government	3 051.452	6.80
	Revenue from investment	200.000	0.44
Total		45 022.953	100.00

Source: General Statistics Office (2015), *Hai Phong Statistics Yearbook.*

Figure 4.3. **Public revenue generated in Hai Phong**

VND billion (2004-2014)

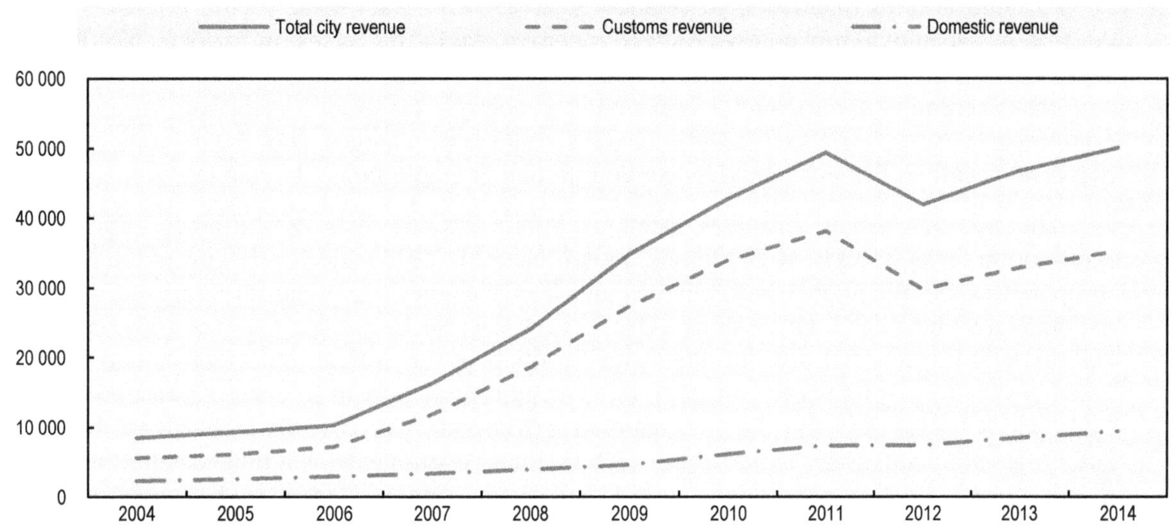

Source: City of Hai Phong (2015), "Answers to the OECD case study questionnaire", internal document, unpublished.

Figure 4.4. **Budget per inhabitant in Hai Phong, Bandung and Bangkok (USD, in purchasing power parity)**

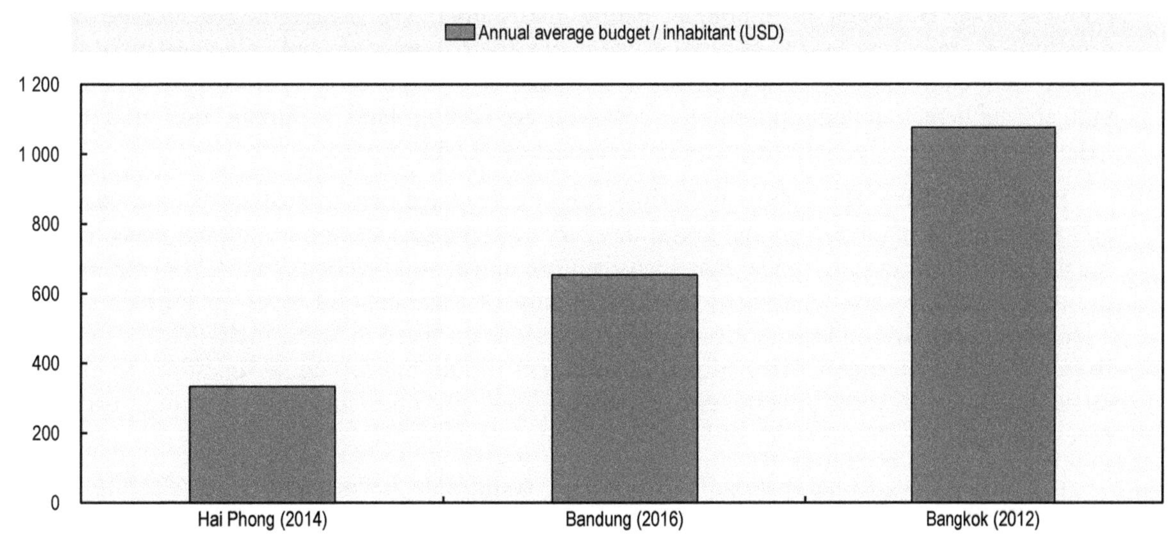

Source: Calculations based on City of Hai Phong (2015), "Answers to the OECD case study questionnaire", internal document, unpublished.

The means to pay for green growth needs to be planned from the outset and included in the green growth plan. However, there is little evidence that this was done in Hai Phong, as its green growth action plan seems to be constrained by the lack of financial resources. To gather the resources for urban infrastructure development, Vietnamese cities use different options. Ho Chi Minh City issues urban bonds with the approval from central government and established the HCMC State Financial Company; Da Nang, on the other hand, develops its infrastructure through initiatives such as the exchange of land for infrastructure and using the land fund for infrastructure development. However, the World Bank has reported that Hai Phong's financial sector is small for a city of its size (World Bank, 2010).

To build a solid foundation to finance green growth, Hai Phong should continue to develop its capacity to rely on **own-source revenues, including taxes and fees to pay for services**. Several revenue options are available for Hai Phong, to explore with the support of the central government: improved property taxation, selective use of non-property taxes, user and benefit charges, road user charges and the reduction of perverse incentives that encourage longer commuting or single-family housing.

The use of charges and fees, such as public parking fees, is extremely limited (accounting for only 0.5% of Hai Phong's total budget in 2013) and barely covers the cost of operation. In OECD countries, charges and fees account on average for 15.2% of sub-national governments' total revenue (OECD, 2015b). Such charges can not only raise Hai Phong's revenue but also promote green growth in the critical opportunity areas for green growth (e.g. solid waste collection fees, wastewater treatment fees, etc.). The Law on Fees and Charges adopted by the national government in 2015, which gives local authorities more autonomy in managing such financial instruments, is a timely opportunity in this regard. Another possibility would be taxing the release of new land, to maximise usage and to help finance green infrastructure development. The revenue from land rent could also be used more flexibly; currently, its use is limited to infrastructure investment. Hai Phong may wish to avoid being in the situation of relying on land sales, due to the number of drawbacks that they entail. For all local governments in the People's Republic of China (China), land leases account for about 30% of revenue; they also represent a low political cost. But even in a unique setting like China, the disadvantages include the sensitivity of land revenues to the real estate cycle; the riskiness of land value collateral for loans; the temptation of "easy money", leading to overspending in local government budgets; underestimating opportunity costs of converting land to urban use; and the exhaustible nature of government land as a resource (OECD, 2015c).

In the long term, Hai Phong authorities might consider a new mix of service provision and financing, to include regional taxes, delivery of some services on a regional basis, and the adoption of a revenue model that focuses more on self-sufficiency. Taxes, tax rates and the object of taxes are determined by the Ministry of Finance, and local governments in Viet Nam do not have the authority to make decisions on these matters, including for taxes from "domestic revenue". The national government should thus continue its efforts to give more autonomy to local governments in raising local revenues that can encourage urban green growth.

Fiscal transfers to sub-national government could also be a useful tool. Despite its severe fiscal condition[5], the national government could consider increasing the share of transfers to the city for specific green growth projects. Viet Nam may use direct transfers earmarked for infrastructure, as India does through the Jawaharlal Nehru Urban Renewal Mission, whose aim is to develop infrastructure in Indian cities. It provides co-financing for cities to improve efficiency in urban infrastructure and service delivery and to promote transparency, accountability, and public participation. The program's success can be attributed to national leadership accompanied by financing, a demand-driven approach, capacity building and peer-to-peer learning in local government (Ellis and Roberts, 2016, Matsumoto and Nuttall, 2014). In this case, it should go to infrastructure that demonstrates its contribution to green growth objectives. Hai Phong has significant needs to support green growth in the port that could be partially taken care of by earmarked national transfers. Such targeted transfers would help concretely to implement the national and local green growth strategies.

Encouraging fiscal decentralisation

To support green growth initiatives in Hai Phong, the government of Viet Nam could enhance the process of **fiscal decentralisation**. This is a challenging task for Viet Nam, but also for many OECD countries. Across the OECD, spending on decentralisation has far outpaced revenue decentralisation. One-third of government spending and two-thirds of public investment is handled by the sub-central level of government, whereas less than 15 percent of tax revenues go to the sub-central level. This is an obstacle to sub-national accountability and good fiscal performance (OECD, 2013a). In addition, the direct lines of "guidance" from each ministry to each city department blur the decision-making process, and reduce the incentives for the city to take initiatives.

In Viet Nam, an increasing proportion of taxes has been assigned to local governments since the 1990s (World Bank, 2010). Since 1996, provincial governments have been given significant powers over a wide range of activities regarding expenditure. In 2002, under the State Budget Law, provincial governments were given more explicit powers to prioritise resources, including determining budget allocations to different sectors and to lower levels of government. The 2002 State Budget Law also allowed provincial governments to borrow for infrastructure investment. However, the central government sets the income cap each year for the city and influences many of the decisions relating to investments, land use and land development. This is in sharp contrast with OECD countries, where most of these decisions are made at the local level. Furthermore, although in many countries municipal governments depend on national/central government transfers for funding, they are free to set fees for public services to increase city tax income. Hai Phong, like the rest of the provincial governments in Viet Nam, depends on approval and resources from the central government to act. This means that the city operates as an administrative agency of central government rather than an autonomous provincial government.

Moreover, the hierarchical fiscal structure of Viet Nam limits the possibility for joint projects between cities within the same region. There is no evidence, for example, of common strategic investment projects between Hai Phong and other cities, such as Ha Noi. Hai Phong's planning and budgeting have to be approved by the central government, and anything that is not in line with central government's priorities or national policies is not approved. Plans and resources for green growth need to be consistent with national priorities if they are to be approved by the central government. This is not bad in itself, because there needs to be consistency between national and local objectives, but to a large extent, this process does not acknowledge that to meet certain targets, cities need to work in partnership, with the possibility of making adjustments to plans, programmes and investment plans, etc., that are particular to their context.

Districts and communes are highly dependent on fiscal transfers to finance their expenditures. Since income is not under the control of the service directors, the level of spending by city departments bears no relationship to demand. Although the intergovernmental fiscal relationships below the provincial level show gradual progress toward fiscal decentralisation, districts and communes rely on fiscal transfers for more than 60% of their fiscal expenditure. It may be argued that all departments are underfinanced, but the centralised controls over policy and priority smother incentives for department heads to innovate, seek greater efficiency and improve productivity.

In addition, districts and communes spend most of their fiscal revenues on personnel costs, which greatly limits their budgetary discretion.

To a large extent, the success of Hai Phong's public finances depends on how the vertical intergovernmental fiscal relations are structured. The experience of OECD countries shows that there is no single magic bullet for reform that is right for all countries (OECD, 2013a; OECD/Korea Institute of Public Finance, 2012). However, the reform process should start at the central level of government, taking a metropolitan view of reform choices. Improving co-ordination across levels of government, as suggested above, would also complement the fiscal decentralisation process. Co-ordination is best made mandatory rather than voluntary, as the experience of India shows (Bahl et al, 2013). In addition, the national government and Hai Phong could consider the OECD Principles for Effective Public Investment Across Levels of Government to strengthen the effectiveness of public investment in green growth.[6] The principles provide useful guidance on how to co-ordinate public investment across levels of government and policies; strengthen capacities for public investment and promote policy learning; and ensure proper framework conditions for public investment at all levels of government.

Expanding public-private partnership in green infrastructure

Attracting and mobilising private finance is a critical complementary strategy the City of Hai Phong can pursue to bridge the financial gap to achieve urban green growth. Capacities to build sound public-private partnerships (PPPs), in particular, should be developed at the national and local levels, in the key opportunity areas mentioned in Chapters 2 and 3. So far, only a few PPP projects have been implemented, mainly in the port (Hai Phong International Gateway Port) and the water and sanitation sector. In Viet Nam in general, very few PPPs have involved foreign investors, because PPP processes are unreliable (ADB, 2012). Recently, however, the government has been developing a new PPP strategy to build credibility in the private sector. The major challenge is to close the financial viability gaps (e.g. low tariffs) in the opportunity areas, in particular transport, solid waste, water and the power sector (ADB, 2012).

Mobilising private finance through the smart use of public finance could help close the green investment gap. This can be done through incentives at the national level, since the Ministry of Finance has the authority to impose taxes, but also by progressively granting local governments more autonomy in creating similar incentives and fiscal regulations, as mentioned above. For instance, the Ministry of Finance of Viet Nam may enable Hai Phong authorities to adopt tax incentives that provide funding or tax reductions for private companies investing in renewable energy, retrofitting buildings or other green projects. The national government and Hai Phong may wish to target the use of public finance to scale up private financial flows into green investment through measures such as guarantees, insurance products and incentives, combined with policy support. According to the World Economic Forum (2013a), there is strong potential for increased lending, advancing and rolling out de-risking instruments, using carbon credit revenues, and targeting grant money combined with technical assistance, to attract much greater private investment. The national government could also help Hai Phong to take similar measures, to strengthen its financial market.

4.4. Enhancing administrative capacity

As in other countries in the region, within Vietnamese sub-national governments, sectoral complementarities between green growth programmes, green growth policies and existing policy instruments are weakened by the limited capacity of the public administration (e.g. low human and financial resources, weak institutions, dependence on higher levels of government). In addition, weak regulatory frameworks do not facilitate cross-sectoral integration.

Public administration capacity challenges in Hai Phong

An inefficient government bureaucracy is one of the factors that most affects the competitiveness of Viet Nam (World Economic Forum, 2013b). Compared to its peers (Ha Noi, Da Nang, Ho Chi Minh City and Can Tho), Hai Phong's public administration has room for improvement.[7] According to the 2015 Provincial Governance and Public Administration Performance Index (PAPI) Report (Box 4.4), it is not a top performer on governance and public administration reform, which is surprising given the economic importance of the city for national development.

More specifically, the 2015 PAPI Report suggests that Hai Phong create more opportunities for every citizen to participate in political activities and policy making, and to raise awareness of grassroots democratic rights. Participation in decision making, when citizens are called upon for voluntary contributions, is low (Figure 4.5). If Hai Phong is to raise residents' awareness and understanding of green growth, and to enlist their co-operation, this will be particularly important. Tackling corruption is another area in which Hai Phong could help to increase its competitiveness. Corruption is the second most problematic factor in Viet Nam in doing business, according to the World Economic Forum (2013b). Irregular payments and bribes, the burden of government regulations and transparency of policy-making are areas where national and sub-national governments have opportunity for improvement. Controlling petty corruption in the provision of public health care and primary education services remains a challenge for city authorities. Citizens are more likely to be asked to pay bribes when applying for land-use rights certificates and applying for jobs in the public sector (Huyen, 2014). Equity in employment presents the greatest challenge, and willingness to fight corruption appears to be more limited in Hai Phong and Ha Noi than in the other three provincial cities.

Hai Phong's strongest areas are public administration procedures and public service delivery, although these also leave room for improvement. Construction permits and certification procedures require streamlining, since it can be too cumbersome for citizens and businesses to comply with regulations (CECODES, VFF-CRT & UNDP, 2015). At the national level, Hai Phong was ranked among a group of poor performers in the 2015 Performance Report. This suggests that the city government faces critical governance and public administration challenges not only in meeting citizens' expectations but also in facilitating green growth initiatives. Encouraging green growth requires private-sector investment, but this will not flow unless Hai Phong provides evidence of strong institutions and a public service that performs at high levels of ethical standards and efficiency.

Box 4.4. **Viet Nam Provincial Governance and Public Administration Performance Index (PAPI)**

The Viet Nam Provincial Governance and Public Administration Performance Index (PAPI) is a policy monitoring tool that reflects citizen experiences with central to local governments in performing their governance, public administration and public service delivery functions. PAPI helps create motivation for public officials, civil servants and public employees at different levels and in different sectors to perform better in public sector management and public service delivery. PAPI ranks the performance of Vietnamese cities based on citizens' answers to a survey. Cities are assessed in six areas: i) participation at local levels; ii) transparency; iii) vertical accountability; iv) control of corruption in the public sector; v) public administration procedures; and vi) public service delivery.

Source: CECODES, VFF-CRT & UNDP (2015), *The Viet Nam Provincial Governance and Public Administration Performance Index (PAPI) 2014: Measuring Citizens' Experiences*, joint Policy Research Paper by Centre for Community Support and Development Studies (CECODES), Centre for Research and Training of the Viet Nam Fatherland Front (VFF-CRT) and the United Nations Development Programme (UNDP), http://papi.vn/documents/provincial-profiles/2014/en/PAPI2014_FinalReport_SmallSize_ENG.pdf.

Figure 4.5. **Governance and public administration performance in Viet Nam's major cities**

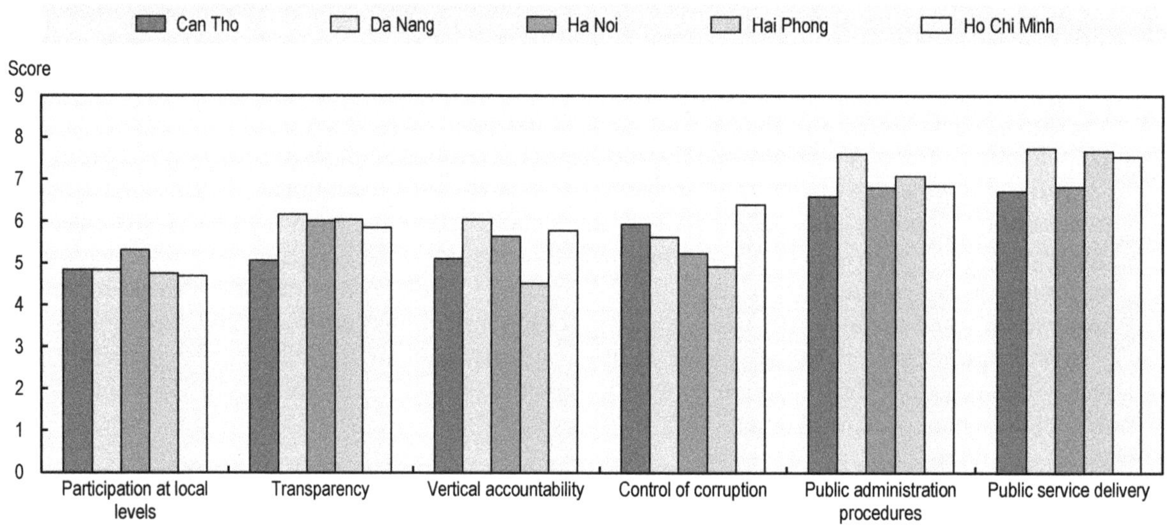

Source: CECODES, VFF-CRT & UNDP (2015), *The Viet Nam Provincial Governance and Public Administration Performance Index (PAPI) 2014: Measuring Citizens' Experiences*, joint Policy Research Paper by Centre for Community Support and Development Studies (CECODES), Centre for Research and Training of the Viet Nam Fatherland Front (VFF-CRT) and the United Nations Development Programme (UNDP), http://papi.vn/documents/provincial-profiles/2014/en/PAPI2014_FinalReport_SmallSize_ENG.pdf.

Strengthening local public administration to underpin green growth strategies

To support the Green Growth Strategy Action Plan, Hai Phong may wish to focus on enhancing the capacities of the local public administration. The objectives should include a skilled and competent public workforce, measures to tackle and prevent corruption, and a clear and sound legal framework.

Hai Phong's public service requires investment in skills and competencies

Hai Phong, like any other level of government in Viet Nam, faces a fundamental challenge: how to attract, retain and motivate civil servants. Attractive opportunities outside the public service lure many qualified and talented people away from the public administration which risks leaving the civil service under-qualified and under-motivated. This could diminish the city government's capacity to implement the plans and programmes for transforming Hai Phong into a green seaport city. Two factors help to explain this: low levels of remuneration and a lack of opportunities and encouragement of talent. Information provided for this case study suggests that high-quality, well-remunerated jobs in the public sector are limited and that the city government lacks specialised personnel with a clear understanding of green growth.

The salary structure for civil servants in Viet Nam includes a range of allowances, bonuses and additional income beyond the basic salary, which accounts for only 30% of total remuneration (World Bank, 2010). The mismatch between public sector and private-sector salaries, which are higher on average (Koh, Du Dam and Nguyen, 2009), has important implications for the civil service in Hai Phong, with the risk of brain drain and the temptation for civil servants to enrich themselves outside of the rules. Higher salaries for vulnerable positions (i.e. highly skilled personnel needed for green growth-related activities, or those in charge of procurement and finance) and making salaries commensurate with risks and with market alternatives could mitigate one factor that contributes to corruption. Increasing salaries across the board, however, is not a realistic option. According to the World Bank (2010), expenditure on salary and wages for government personnel already accounts for nearly half of the recurrent government expenditure. One way forward would be for Hai Phong, with the support of the central government, to simplify the salary scales, by targeting increases on the skills profiles for which salaries are most out of line, or by matching salary increases with a reduction in the number of state employees through strategic workforce planning. Making salary increases selective, and matching them to improved qualifications, skills or responsibilities rather than length of service, could have a positive effect on motivation and recognition. The central government could explore full delegation of pay settings to local governments, so they could adapt them to their needs. This process could begin with a pilot test in the provincial cities.

The process of organising, assigning jobs to and promoting civil servants could be refined. This is a concern for all levels of government. In several organisations, many good staff have not been encouraged and assigned appropriate tasks, and many unqualified staff are not replaced, leading to understaffing of competent personnel. In some cases, staff allocation is not tailored to the structure of administrative departments. The Law on Cadres and Civil Servants has adopted the principle of meritocracy, but translating it into concrete recruitment and managerial processes is a challenge for Hai Phong authorities. Interviews for this case study indicated that although entry-level recruitment is based on competitive examinations and personal interviews, much emphasis is placed on theory and knowledge, with relatively little attention to experience, competencies, skills or appropriateness of those skills for the position. Assessment of individuals' drive for results is weak. Entry-level recruitment suffers from untrained interviewers and opaque evaluation criteria and recruitment results, creating extra-legal opportunities for recruitment. The recruitment of non-entry level positions is based on grade promotion examinations, but the merit orientation is undermined by the lack of competition from external candidates, a common weakness in career-based systems such as Viet Nam's (OECD, 2005). For non-entry positions, greater competition and transparency, and more openness to external competition, could increase the pool of applicants and help to bring in fresh ideas from outside the public service (World Bank, 2010).

To promote merit in managing the public workforce, Hai Phong authorities require good information on civil servants' performance and links to career success. The current methods for assessing performance do not link performance and ratings, and performance appraisals are not strongly used as a tool for rewarding good performance through promotion and appointment, or for addressing poor performance. This could be improved by making evaluations more objective. The performance evaluation system could be improved by developing job descriptions and competence frameworks (OECD, 2005, 2008a, 2011). Hai Phong authorities may wish to introduce a performance management system that includes elements such as: planning work and setting expectations, continually monitoring and appraising performance, developing the capacity to perform, periodically rating performance and rewarding good performance (OECD, 2008a).

Training and upgrading of civil servants in Hai Phong, as in the rest of Viet Nam, is only in the initial stages. Sometimes the content is not fully relevant to the civil servants' job. The curricula are not practical, and much emphasis is placed on theories and too little on the training of skills relevant to the job. Staff performance improves only a little after training. The requirement to obtain all formal qualifications (foreign language, computer skills, political training, etc.) has led to a situation where staff try to become eligible to take recruitment or promotion exams. However, even though they have formal qualifications, they may not be competent for the task. Recruitment and promotion should focus more on skills and competencies than formal qualifications. A certification of competencies could be one way of ensuring that all experience acquired is being put into practice and acknowledged.

Box 4.5. **Elements of a performance management system**

A performance management system is a systematic process by which an organisation involves its employees, as individuals and members of a group, in improving organisational effectiveness in the accomplishment of an agency's mission and goals. It should include elements for:

- planning work and setting expectations
- continually monitoring and appraising performance
- developing the capacity to perform
- periodically rating performance in an adapted fashion
- rewarding good performance.

Performance management systems should link the management of people to the institutional goals and strategies.

Source: OECD (2008a), *The State of the Public Service*, http://dx.doi.org/10.1787/9789264047990-en.

Encouraging green growth requires high ethical standards in the public service

In Hai Phong, as noted above, dealing with corruption and misconduct in the public service remains a challenge, but this is a nation-wide problem. Misuse of public resources, in particular, should be avoided (World Bank, 2010). Complicated administrative procedures that lack clarity, excessive regulations, the opaque nature of decision making, lack of public information, bureaucratic discretion on the part of middle-level officials, long delays and low remuneration levels for public servants all typically result in the misuse of public resources. Large-scale infrastructure investment may be prone to corruption and mismanagement, given the degree of public officials' discretion over investment decisions, the many phases and large sums of money and stakeholders involved, including elected and non-elected public officials, lobbyists, civil society organisations, trade unions, regulators, contractors, consultants, engineers and suppliers. The project design phase and the implementation phase of such investments often involve corruption in the form of bribery, favouritism towards certain bidders, conflicts of interest, fraud or collusion. Corruption and mismanagement Hai Phong come at a high cost. Direct costs include bribe transfers, higher expenses, scarcity of essential services, lower quality and misallocation of public funds. One of the common consequences is the purchase of lower-quality material to compensate for the section of the budget used to bribe officials. Construction permits are sometimes given to build in areas rich in natural diversity.

Considering the need for substantial infrastructure investment in rapidly urbanising cities, Vietnamese authorities at all levels of government would need to act to prevent corruption and mismanagement by strengthening the legal framework; for instance, Mexico has created the Federal Anti-Corruption Law on Public Procurement to address and prevent corruption and fraud. Viet Nam, and Hai Phong in particular, could promote ethics and integrity in the public service by aligning those efforts with the OECD Principles for Managing Ethics in the Public Service (OECD, 1998) and use the OECD Guidelines for Managing Conflicts of Interest in the Public Service. At the national level, the government and Vietnamese Communist Party emphasise their opposition to public sector corruption and have issued decrees on corruption and on the elimination of wasteful practices in the public sector. To support these efforts, the creation of a Commission for Public Ethics like Brazil's, responsible for implementing the codes of conduct as well as for oversight and evaluation, would be of great relevance (OECD, 2010a). An alternative would be to adopt measures to increase the openness and transparency of the civil service to strengthen ethics management. These measures will require strong leadership and political backing at both the local and national level.

Improving regulation and dealing with red tape could facilitate investment in green growth

As mentioned above, one of the barriers to greater public sector performance in Hai Phong and one of the sources of misconduct is an excess of complex, confusing regulation. Land-use certification has become a slow process, because of land abuse or misuse (i.e. houses built on agricultural land), ownership issues and other disputes arise, and the documentation required is so onerous, it is regularly submitted incomplete. The biggest challenge is typically to determine land origin and the length of land use (Konrad-Adenauer Foundation, and the Association of Cities of Viet Nam, 2009). To promote green growth investment, it may be necessary to improve the quality of regulation. The aim should be to reduce regulatory complexity and uncertainty, and cut red tape, reducing unnecessary burdens created by bureaucracy and paperwork. In doing so, Hai Phong authorities may wish to follow the OECD Guiding Principles for Regulatory Quality and Performance, which suggest a number of recommendations to make regulations more responsive and predictable.[8]

It is clear that Hai Phong's government will not be able to start such a process without central government support, as local regulation must be in line with the national regulatory framework. Nevertheless, Hai Phong may still have some room for manoeuvre in designing secondary legislation, in order to encourage an environment favourable to green growth. One aspect that should be considered carefully is to review regulation considering the principles of good regulation and from the point of view of those affected rather than of the regulator. Hai Phong authorities could consider alternatives to regulation where appropriate and possible, including self-regulation, giving greater scope to citizens and firms. In analysing such alternatives, their cost, benefits, impact on competition, market openness and administrative requirements must be taken into account. Hai Phong's authorities could ensure that regulations, regulatory institutions and regulatory processes are transparent and non-discriminatory. The aim should be to ensure that firms in an industry are not subject to firm-specific benefits or costs arising from regulation. The central government could support this process of regulatory reform by encouraging development management capacity and performance at sub-national levels of government. The objective should be to reduce regulatory costs and barriers at the local and regional level that limit competition and impede investment, business growth and job creation. The central government could also facilitate local variations and experimentation in regulatory approaches when it is nationally beneficial. Hai Phong may need to look for national and international support for capacity-building on regulatory management. Attention may need to be paid to the promotion of e-government and administrative simplification when appropriate. In turn, Hai Phong could also promote information sharing and transparency with lower levels of government to overcome asymmetries of information and promote complementarities across regulations.

4.5. Shaping urban green growth from the citizens' perspective

In Viet Nam, as in many OECD member countries, government finds itself under pressure to ensure that public policies are effective and legitimate. Similarly, Hai Phong needs efficient mechanisms to encourage the community's participation in the decisions that affect them, as well as the need to exercise effective oversight of expenditure.

Ensuring citizen involvement in defining urban green growth priorities

Citizen participation in public affairs, such as in the decision-making process determining provincial budgets or urban planning, has been minimal. In urban construction planning, the People's Committee of Cities, Districts and Towns typically outsource land-use planning to provincial offices with qualified staff. These plans tend to be based on politically generated targets rather than on local demand or spatial constraints. There is no involvement from citizens; they are not informed of the plans even after they are approved (Konrad Adenauer Foundation and the Association of Cities of Viet Nam, 2009). Another aspect that discourages citizen participation is the format of documents, which are difficult for lay persons to understand. In many cases, people are informed after the fact, given conflicting information about what is going to happen and are instructed that they have no choice in the matter (Konrad Adenauer Foundation and the Association of Cities of Viet Nam, 2009).

According to the PAPI Report 2015, citizen participation in public affairs in Hai Phong and the quality of civic knowledge has decreased since 2012. The city also faces the challenge of adopting effective mechanisms to engage citizens in monitoring the performance of local governments. Hai Phong needs to improve its accountability to citizens. Regular or *ad hoc* interactions with local authorities or grassroots institutions like the People's Inspection Boards (PIB) and the Community Investment Supervision Boards (CISB) need to be increased, since few citizens are aware of their existence. Grassroots public officials do not seem to play an important role in city dwellers' life, as few citizens contact them about problems concerning their families, community or local governments. The governance arrangements may also need to be reviewed carefully to ensure their effectiveness. For example, when the Hai Phong government interacts with its citizens in broad consultations over investment options, budget constraints are often not taken into account. This means that neither policy makers nor citizens consider the possible trade-offs within the limited budget in broad consultative fora. As a result, spending priorities (services and investments) are determined without reference to citizens' needs, as the overall level of spending is determined by budget limits based on expected income, also without reference to citizens' preferences.

Some studies suggest that in the context of rapid urbanisation in Viet Nam, citizen participation is weaker in large cities than in small ones (Nguyen et al., 2015). Promoting authentic participation is particularly challenging in the urban context, because cities normally have populations that are more diverse and face more complex development problems than rural areas (Oliver, 2000). The problem is that Hai Phong, like other provincial governments, is subject to forms of upward accountability in its budget decisions. There are certain quotas and standards determined by central-level authorities that must be applied by provincial governments when they allocate their budgets, to ensure that certain development goals are met. This limits the extent to which Hai Phong's plans and budgets can respond to local needs and green growth goals.

According to the World Bank (2010) the urban planning process in provinces in Viet Nam is somewhat participatory, although the vision of full participation may not be reaching its potential. Interviews for this case study suggested that local people rarely participate actively in the urban planning process. The 2009 Law on Urban Planning, for example, requires that the People's Committees of provinces and centrally run cities, like Hai Phong, collect comments of concerned agencies, organisations, individuals and communities on urban planning tasks and urban plans. Consultation could take place via dossiers, conferences, workshops or survey card questionnaires. The time limit for comments is at least 15 days for agencies and 30 days for organisations, individuals and communities. Hai Phong, like many other cities in Viet Nam, may transfer information to block or village leaders, but the transfer of information from block or village leaders to the citizens is less consistent. There is no effective system to gather citizens' feedback or comments to improve urban development plans, or administrative service delivery. Moreover, urban planning and management tasks fall to civil servants, who lack professional training in these fields.

The targets set in the urban plans tend to be handed down from the central level to the provincial level, and from there to communes and districts successively, and are not discussed at the local level or based on realistic analysis and evidence. The targets set in the Green Growth Promotion Plan of Hai Phong were not widely discussed with all stakeholders involved. Public awareness of citizens' responsibility for environmental protection is low. The lack of information leads to confusion, misunderstandings and mismanagement of resources. One key drawback in the green growth strategy is that many stakeholders, including the private sector and citizens, have little understanding of the concept of green growth, and no information on how they can contribute to it. This suggests that consultation with the public is limited, and undertaken only in areas considered to significantly affect the public. This lack of participation in the planning process may be explained by institutional factors. Planning agencies are well set up at national and provincial-city level, but not at the levels of districts and communes. The World Bank (2010) found that in Viet Nam, time constraints, shortage of skills and a lack of full-time planning staff have all contributed to a lack of participatory planning at the grassroots level.

Strengthening the relations between Hai Phong government and citizens

To encourage citizen participation in urban planning, environmental management and decision-making on green growth, Hai Phong needs to create or reinforce formal and informal communication mechanisms, to encourage information sharing, collaboration and co-operation among civil society groups, levels of government and between levels of government and civil society. Hai Phong may benefit from the experience of Kitakyushu in Japan, where citizens, notably women's associations, are taking a greater part in promoting green growth (Box 4.6). Montreal, in Canada, also demonstrates how green growth projects can be transformed into genuine social projects. The city of Montreal is promoting an ecological transition of the economy. To this end, the Transition Montreal 21 Programme aims to transform environmental liabilities into productive assets, while generating business opportunities. Montreal's authorities acknowledge that the ecological transition can only succeed with the help and support of citizens. During the implementation of the initiatives, comprehensive public consultation is carried out to ensure that this is a social project.

Hai Phong could go the extra mile by not just seeking feedback but collaboration. Viet Nam has a long tradition of working at grassroots level. Based on this experience, Hai Phong could promote the creation of citizens' organisations promoting green growth and sustainable urban development. The Friends of Midcoast Maine in the United States offers one example of how such groups can promote grassroots planning that gives voice to citizens. Such groups educate and advocate for healthy downtowns and the preservation of the rural landscapes, and organise conferences and meetings that involve citizens in grassroots planning (Box 4.6).

It is recommended that Hai Phong create, expand and implement legal and regulatory frameworks that ensure citizen participation in sustainable development debates including green growth objectives. Croatia, for example, has developed a law requiring consultation in legislation and regulation that recognises that the results are part of the decision-making process.[9] In addition, Hai Phong could focus on procedures and techniques for civil servants to involve people in the early stages of planning, in particular those related to urban development and green growth, before decisions on how to proceed are taken. Developing communication at both the ward and city levels is necessary, particularly in the areas of discussion and consultation. The city government could develop guidelines recommending which kinds of information methods are appropriate. Participative techniques could be taught to civil servants, focusing on improving communications to increase public awareness. Such skills would include skills for presentation and running meetings, developing quality feedback forms, using meta-planning to gather and structure citizen feedback, etc. One other key aspect would be to develop and strengthen the capacity of individuals to participate in green growth and sustainable development decision-making, with an increased base of knowledge of green growth and public participation practices. In this sense, formal and non-formal education and training programmes are essential for effective public involvement in decision-making.

Box 4.6. **Partnering with citizens for green growth – Kitakyushu and Maine**

Environmental governance in **Kitakyushu**, Japan, has a long tradition of citizen participation. Citizen movements, notably women's associations, were at the forefront of pushing for environmental improvements in the 1960s to 1980s. More recently, citizens played an important role in conceiving sustainability plans and documents such as the Grand Design and the Green Frontier plan, through participation in a large number of town-hall meetings. An important recent initiative is the Kitakyushu Clean-Up Union, organised by a community-based sanitation and environmental association, with over 70% over Kitakyushu's households participating. This association includes a large number of elderly people who aim to pass on their experience and know-how to younger generations.

In the state of **Maine**, in the United States, a local non-profit organisation, "Friends of Midcoast Maine", is an example of grassroots planning. It aims to help Midcoast communities plan for a vibrant and sustainable future. It provides: i) education, workshops and technical assistance; ii) project endorsements; and iii) advocacy for sound planning and sensible growth. It provides assistance to communities based on smart-growth principles such as: i) encouraging the development of compact towns and centres separated by rural countryside; ii) directing public funding to guarantee investment in existing schools, neighbourhoods and infrastructure in compact mixed-used centres; iii) protecting the quality of the environment, open space, scenic landscapes and historic features; iv) providing a range of housing alternatives in fast-growing communities; v) ensuring stakeholder environment and regional collaboration on land-use and transport planning; vi) promoting transport choices that reduce auto dependency; and vii) supporting economic vitality by encouraging business enterprises in downtowns and villages, especially locally owned business.

Source: OECD (2013b), Green Growth in Kitakyushu, Japan, http://dx.doi.org/10.1787/9789264195134-en; Friends of Midcoast Maine (n.d.), *Friends of Midcoast Maine – A grassroots initiative for sensible growth*, https://lcrpc.org/uploads/visual_edit/preservingsmalltowncharacter-1.pdf (accessed 10 November 2016).

Enhancing citizens' access to information

Setting a framework for dialogues between government and citizens will help to rebuild relations of trust and confidence on which democracies depend and which legitimise decisive and effective action by government. Such a framework builds on the legal rights of citizens to information, consultation and active participation on government policies and the institutions charged with the tasks. Much of the legal and policy framework concerns access to information, the basis for strengthening government-citizens relations. Some countries, such as Finland, Mexico, Sweden and the United States, have passed legislation on citizens' access to information. Additional provisions enhancing access to information are made in laws in areas such as the environment, consumer protection and health. This framework should define or include: sound legislation, clear institutional mechanisms and an independent judiciary for enforcement. In building such a framework, Hai Phong may wish to keep two things in mind. First, it is necessary to define what information is accessible to citizens, what the exceptions are, whose information is available and in what form, and data protection issues. The second is the need to define how information can be accessed and received, including for example, maximum response time, what to publish actively, and how to convey and appeal a refusal.

A careful selection of tools for access to information and consultation is needed...

In developing an overview of possible tools for access to information and consultation activities, Hai Phong authorities may consider the different characteristics of tools in terms of form, support and channels of delivery; the different approaches of government to use the tools; and the possibility of using a mix of tools. The city government can passively provide access to information, or inform citizens actively, on its own initiative. It can use information products, direct and controlled channels or independent channels to deliver the information. Interfaces for citizen access; catalogues, registers and indexes; official documents; thematic reports; and handbooks, guides, brochures and leaflets are just some of the tools government can use to disseminate information on green growth. For information, Hai Phong authorities need to make sure to adapt the tools to suit the objectives, since they have to be used differently depending on the targeted goals. They also need to select the tools to suit the intended audience, in terms of language, style and presentation. This could be used for example to explain to a wider audience what green growth is about. Government also needs to ensure that information reaches the public and that information is attractive so that it can be used by citizens. It is in the government's interests to ensure that the media and citizens receive information that is as complete as possible.

Hai Phong's government could extend its effort to receive input from citizens, whether unsolicited or by inviting citizens to respond. Unsolicited feedback may contain valuable information. Tools to support the use of unsolicited feedback include letterboxes, information management software and analytical reports. The government may also receive solicited feedback from citizens through questioning, listening and reporting; comment periods and actions; focus groups; surveys; and public opinion polls. Hai Phong authorities may also use some tools for *ad hoc* consultation, such as the inclusion of citizens in consultative bodies; workshops, seminars, conferences; public hearings; non-binding referenda; open hours; citizens' panels; citizen report cards and advisory committees. It is important that government inform citizens about the when, where and what of the consultation before it starts. Who to select for consultation is not only crucial for the quality of the answers received through consultation, it is also crucial for the effectiveness of the consultation. Setting, publishing and following clear rules and conducting selection transparently can help to increase trust and effectiveness. Government also needs to use the information received to avoid damaging its relations with citizens. Finally, the use of social media and other platforms can help Hai Phong's government listen to, engage and be responsive to the needs of residents. This holds great potential that should be exploited in building dialogue with citizens. Australia, for instance, has developed guidance on online consultation and the use of social media, to promote participation in decision-making processes and to complement traditional consultation methods.[10]

... as well as for disseminating green growth information in civil society

To achieve green growth objectives, Hai Phong needs to increase citizens' awareness of green growth and environmental protection. One crucial element is the transmission of data and information. It may be useful for the city government to state clearly its understanding of what green growth involves, to make sure all citizens have the same shared vision and information. Green growth information could be disseminated by different means: manuals and guidelines, thematic reports, conference proceedings, data yearbooks and state of the green growth reports. Hai Phong authorities can also make more use of the Internet as an instrument of green growth information dissemination. The Internet makes it possible to share large amounts of information through periodical bulletins, databases with connected cartography and general services, including information meant for businesses and other administrative bodies and users.

Ensuring that the information disseminated is well grounded in technical-scientific terms is of key importance in helping citizens to understand green growth issues. Investing in continuous education and professional training is another way of raising awareness. Hai Phong authorities could promote a variety of educational activities and training courses on green growth. Italy's UNESCO Commission organises an annual National Week for Education in Sustainable Development, and awards the DESS logo to all initiatives that help achieve the objectives of the decade. Such training activities can also provide opportunities for contact and exchanges in the field of green growth and result in national or international co-operation initiatives. In Denmark, the Danish Board of Technology Foundation was set up by the Danish Parliament to disseminate knowledge about technology. It also established the Centre for Collaborative Democracy to strengthen democratic dialogue and co-operation among citizens, elected officials, experts, business and organisations, and offers advice on planning and executing participatory processes.[11]

4.6. International co-operation for green growth in Hai Phong

Prioritising official development finance to contribute to urban green growth more comprehensively

International co-operation plays a critical role in Viet Nam's development. One of the forms it can take is official development finance, which includes **official development assistance (ODA)** and non-concessional **other official flows (OOF)**. Viet Nam and Hai Phong have benefited from significant amounts of official development finance. Between 2002 and 2014, Viet Nam has received around USD 68 billion in the forms of ODA and OOF,[12] and ranks as the second top recipient of official development finance in Southeast Asia, after Indonesia. It received more than twice the amount of ODA pledged to the Philippines in the same period, and more than seven times the amount received by Thailand (Figure 4.6).

Figure 4.6. **Total official development finance committed to selected ASEAN countries**

USD billion, in constant 2013 prices (2002-14)

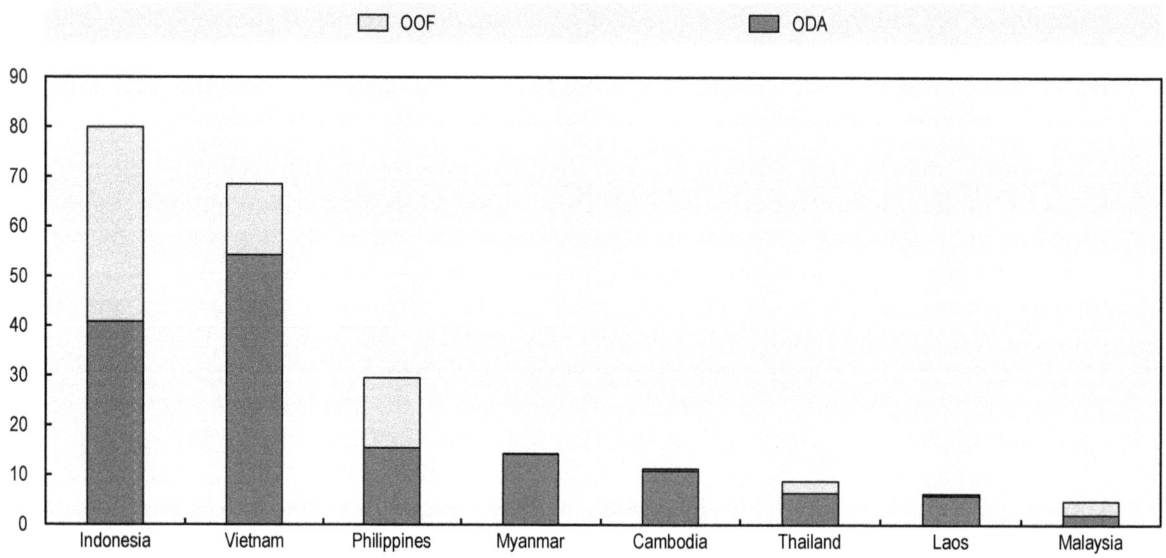

Source: OECD (2016), "Creditor Reporting System: Aid activities targeting Global Environmental Objectives", *OECD International Development Statistics* (database), http://dx.doi.org/10.1787/9c778247-en (accessed 12 July 2016).

Not all official development finance committed to Viet Nam benefits (urban) green growth. Only USD 20.3 billion (i.e. 30% of total official development finance) – pledged to a total of 3 963 projects – targeted the environment, climate change mitigation, climate change adaption, biodiversity and desertification (i.e. the goals set by the Rio Conventions).[13] Of these, 410 projects totalling USD 8 billion (i.e. 40% of total environment-related official development finance) were committed to Vietnamese cities, including USD 496.7 million for the City of Hai Phong alone (i.e. 2.4% of all environment-related finance committed to Viet Nam) (Figure 4.7).[14] The Government of Japan is by far the highest contributor (USD 432.5 million). The environmental-related international grants and loans committed to Hai Phong between 2002 and 2014 mainly targeted the water supply and sanitation and port development sectors (49% and 43% of all funding to Hai Phong, respectively), more precisely the development of drainage, sewage, solid waste collection and treatment systems, and the construction of infrastructure for the new international port of Lach Huyen (Figure 4.8). By contrast, Bangkok received 95% of its environment-related ODA (USD 2.2 billion) for transport development projects (OECD, 2015d). Most secondary cities in Southeast Asia tend to receive much lower amounts of official development finance than Ha Noi and Ho Chi Minh City. The amounts received by Hai Phong can even be considered relatively high: Bandung, Indonesia, for instance, received only USD 177 million over the same period, despite its comparable population (OECD, forthcoming). The central government's involvement in modernising the Port of Hai Phong, within the framework of the Viet Nam Seaport Master Plan 2020, was a decisive advantage.

Figure 4.7. **Number of official development activities (left) and funding (right) for urban and rural areas in Viet Nam for environmental purposes (including Rio Conventions) (2002-14)**

USD billion, in constant 2013 prices

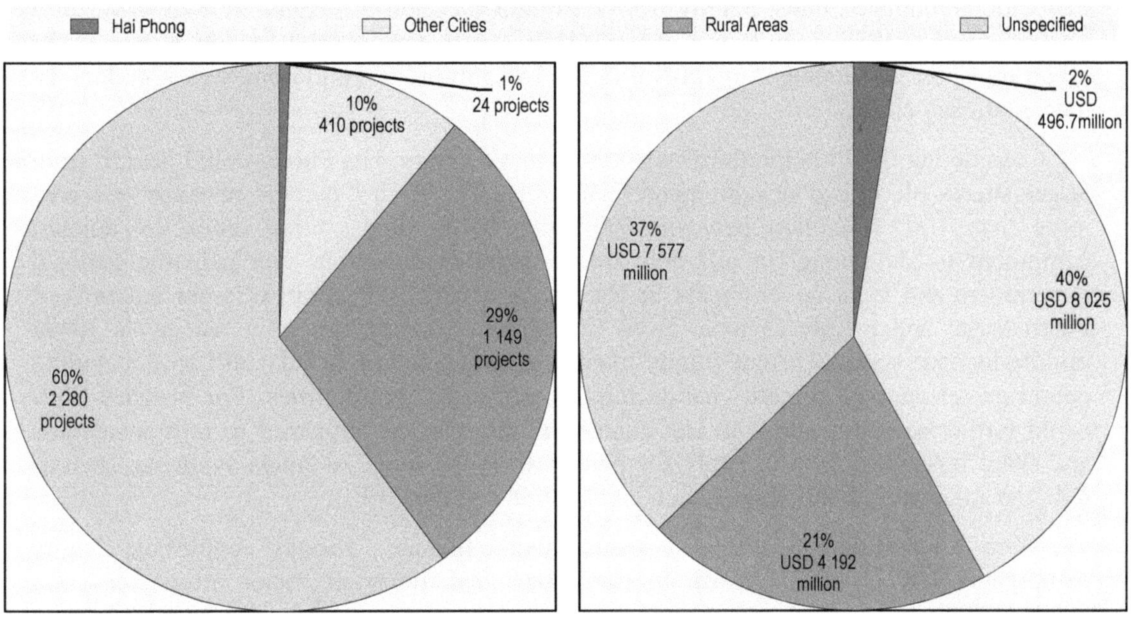

Source: OECD (2016), "Creditor Reporting System: Aid activities targeting Global Environmental Objectives", *OECD International Development Statistics* (database), http://dx.doi.org/10.1787/9c778247-en (accessed 12 July 2016).

Figure 4.8. **Number of ODA activities (left) and ODA funding (right) for Hai Phong for environmental purposes (including Rio Conventions) (2002-13)**

USD billion, in constant 2013 prices

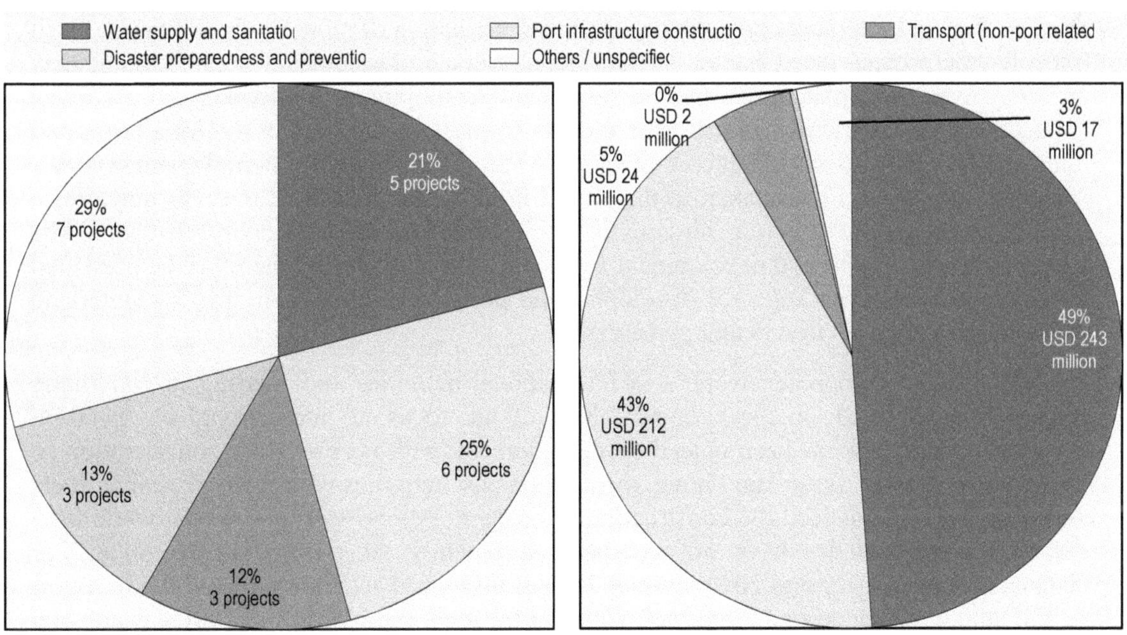

Note: Two projects under "Port Management" included the construction of roads and bridges to connect the new site to the mainland.

Source: OECD (2016), "Creditor Reporting System: Aid activities targeting Global Environmental Objectives", *OECD International Development Statistics* (database), http://dx.doi.org/10.1787/9c778247-en (accessed 12 July 2016).

Data on official development finance presented above include climate funds. Some of them are considered to be ODA, such as the Global Environment Facility (GEF), while others, such as the Clean Technology Fund (CTF), are additional. Since 2003, a total of 32 international climate grants, mainly from GEF and CTF, were pledged to Viet Nam, for a total of around USD 270 million.[15] Among them, ten were destined for cities, mostly for urban transport development in Ha Noi (USD 167.5 million). Hai Phong was not amongst the beneficiary cities.[16]

Considering that it is the third-largest city in Viet Nam, Hai Phong could benefit from larger shares of official development finance. Ho Chi Minh City, for instance, received more than USD 1.3 billion between 2002 and 2013, almost three times the amount committed to Hai Phong. In addition, international organisations can help the national government and local governments in Hai Phong access and mobilise larger amounts of international and private finance. Even though the country has now reached a lower middle-income status, current trends toward enlargement of official aid to developing countries relevant to climate change might open new opportunities. For instance, they could further communicate with Hai Phong and the relevant ministries to raise awareness and share lessons on how to apply for finance from the range of funds available, such as CIF, the Adaptation Fund, the Green Climate Fund and the GEF.

More importantly, official development finance should contribute more comprehensively to green growth in Hai Phong. The transport sector offers one good example. Only two projects contributed to transport development in the city, and focused on the construction of roads and bridges within the framework of port development. No funding was committed to develop complementary and low-carbon transport modes. The construction of the international port of Lach Huyen will generate a surge in cargo volumes in the Port of Hai Phong, and will therefore require higher capacity and more efficient hinterland connections to move goods and containers from the port to warehouses and dry ports, and from the hinterland to the port. The Port of Hai Phong already suffers from inefficient hinterland connections, and the construction of more roads and bridges will not solve these issues (see Chapter 3). The increase in cargo volumes generated by the new port may even exacerbate them. Only a significant development of railways and waterway connections can efficiently increase the competitiveness of the port in the long term, and multi-modal transport should therefore be viewed as the necessary complement to other infrastructure projects undertaken in the city. Similarly, the growth of the port may create environmental externalities (e.g. air emissions, oil spills, loss of biodiversity) that should be addressed early on. Official development finance for the port should therefore be presented as a comprehensive package for developing the port's activities, including its hinterland connections and its environmental performance.

The lack of comprehensiveness in international activities and funding in Hai Phong reflects broader trends in the country. Much of the focus of international co-operation projects in Viet Nam has been placed on development, with no particular consideration so far for green growth. As in Hai Phong, most of the past activities with international partners on transport have financed the construction of bridges, expressways and roads, while low-carbon transport modes have not yet been sufficiently taken into consideration. For instance, between 2010 and 2012, around 25% of ODA in Viet Nam targeted the transport sector, among which around 70% was disbursed for road improvements (Darko, Manh Hai and Whitley, 2015).

Establishing a local investment fund for urban development

Involving local government in application processes or project discussions early on could be an efficient way of ensuring that official development finance activities are more comprehensive. The national government is the main counterpart of international donors and also has authority in the Port of Hai Phong, which may explain why some projects targeted port construction without addressing complementary aspects such as local environmental preservation and hinterland connections. Local environmental degradation and congestion fall mainly on the local government, while the national government gets economic benefit from increased cargo volumes and the enhanced competitiveness of the port. Involving the City of Hai Phong (and port authorities) in such discussions would help clarify the challenges ahead for port development and urban green growth generally.
If the local government is to benefit from the range of international climate funds, it will have to co-operate with the national government, which usually decides how to apply for them and which proposal to submit. Local authorities and the national government could work hand in hand on proposals, in order to optimise their chances of getting financial and technical support.

One potentially interesting option would be to set up **a local investment fund for urban development** in Hai Phong. This could be an opportunity to attract international public finance directly to the city and to give the local government a greater role in co-ordinating international finance to ensure comprehensive development. Local mechanisms of this kind have already been set up in Ho Chi Minh City and a few other cities, such as Can Tho and Da Nang. The French Development Agency has been the main international donor providing support to these local investment funds, and has channelled funds for projects in the health, education, housing, urban development and environment sectors.[17] For some projects, however, these local investment funds will need to collaborate effectively with the national government. Port development (for which the national government is responsible) and railway development (at a scale that exceeds the city's jurisdiction), for instance, fall largely under the jurisdiction of the state but have a strong local dimension.

The Viet Nam Development Bank's role in encouraging urban green growth

The national government in Viet Nam has a key role to play in ensuring that international funding contributes to urban green growth. The Viet Nam Development Bank (VDB) was established in 2006 to administer national development investment and export credit policies, as regulated by the government. One of its functions is to manage and mobilise funds from foreign institutions, including ODA (Box 4.7). As of December 2011, it managed around 60% of total ODA funding sources in Viet Nam.[18] The national government could increase the VDB's portfolio and align its mandate and ODA activities with green growth. VDB should also make sure local investment funds help co-operate on urban development, if such a fund is created in Hai Phong. This could help ensure that a project like port development, initiated by the national government, is coupled with projects managed by the local investment fund and therefore contributes to green growth.

VDB could also assume the functions of a green investment bank, such as the UK Green Investment Bank. These are domestically focused public institutions that use limited public capital to leverage or "crowd-in" private capital, including from institutional investors, for low-carbon and climate-resilient infrastructure investment. This would open a new channel for foreign investors and international organisations to finance urban green growth projects in Hai Phong and other cities in Viet Nam – not directly but through the bank. Green investment banking offers potential for developing the renewable energy and energy efficiency required for the city's growth. The recent adoption of a green growth agenda at the national level offers the opportunity to undertake such a reform and to ensure that international finance contributes to urban green growth. Additionally, the VDB could take on the responsibility of linking up international public finance, foreign private finance and domestic private finance, which tend to be managed separately (Darko, Manh Hai and Whitley, 2015). Bridging this gap could help to leverage significant amounts of funding for the development of green infrastructure.

International partners could also support the urban green growth agenda in other ways, not simply by directly financing large infrastructure projects. They could also help the government develop incentives for domestic actors to invest in green growth projects, such as rail and water transport (Darko, Manh Hai and Whitley, 2015).

Box 4.7. **The Viet Nam Development Bank**

The Viet Nam Development Bank (VDB) was set up in 2006 under Decision No. 108/2006/QD-TTg of the prime minister, based on the reorganisation of the Development Assistance Fund, to implement investment credit policies and export credit policies. VDB is non-profit and wholly owned by the government. It is recognised as a legal entity with chartered capital of VND 15 trillion (USD 75 million), expected to increase to VND 30 trillion (USD 1.5 billion) by 2020.

VDB provides finance for projects and development programmes, as decided by the prime minister, upon economic demand. VDB's functions and mission involve implementing state policies on investment and export credit in line with government regulations; managing the on-lending ODA funds delegated by the government; receiving trust funds for grants and other grant activities; providing finance for small and medium enterprises (SMEs); mobilising funds from domestic and foreign institutions; and performing other duties assigned by the prime minister. Areas in which VDB is actively involved include infrastructure, education, health care and energy. In addition, VDB supports SME with loans from commercial banks under guarantee schemes, to facilitate access to loans.

In terms of international co-operation, VDB has built relationships with multilateral and bilateral financial institutions such as: ADB, the World Bank, US Exim, KFW, Citi Bank, Kexim, DBK, CDB, China Exim, etc. It is also a member of international financial organisations such as the Association of Development Financing Institutions in Asia and the Pacific (ADFIAP) and the Asia-Europe Business Forum (AEBF). VDB is currently working with international financial organisations to implement large-scale projects in Viet Nam, such as the Ha Noi-Hai Phong highway project and the Bac Lieu wind power project.

Source: Asian Exim Banks website, http://www.asianeximbanks.org/content/institutions/vietnam-development-bank (accessed 3 August 2016).

Strengthening international co-operation

Hai Phong already has a network of international partners that provide financial and technical support. **City-to-city co-operation** provides direct and immediate benefits and should be developed more systematically. Co-operation with the cities of Kitakyushu (Japan) on the Green Growth Promotion Plan and Fukuoka (Japan) on wastewater treatment has shown that knowledge transfer, combined with technical and financial support, can make a significant contribution to urban green growth. Both cities signed a Memorandum of Understanding in 2009 to promote international co-operation and exchange, leading to technical co-operation on clean water management, industrial support, business co-operation, capacity building and cultural exchanges. One of the outcomes of this co-operation was the publication in 2015 of the

Green Growth Promotion Plan of the City of Hai Phong, which provides specific policy recommendations for implementing the Hai Phong Green Growth Strategy Action Plan. In April 2014, Hai Phong and Kitakyushu officially became sister cities. This partnership helped to develop green infrastructure and strategies in Hai Phong, while improving local understanding of urban green growth. However, one stand-alone co-operation agreement cannot cover all the infrastructure needs and costs that Hai Phong faces. The local government should look for additional international partners in the Asia-Pacific region and beyond, to develop some of the most urgent infrastructure and capacity needs identified in the Promotion Plan and the present report. Many other port cities have developed efficient green port strategies, with thorough environmental monitoring systems and diverse policies to reduce the port's environmental impact. This could be the object of new city-to-city partnerships.

Another promising initiative to strengthen international co-operation for urban development in Viet Nam, and which could benefit Hai Phong, is the Viet Nam Urban Forum (VUF), chaired by the Ministry of Construction, which brings together Vietnamese government agencies (including cities), international organisations such as UN-HABITAT, academics, cities and NGOs to discuss urban development every year (Box 4.8). The VUF has difficulty ensuring result-oriented outcomes, given the diversity and number of the participating organisations. As a result, it may not offer concrete benefits to Hai Phong, one of its many members. One option worth considering would be city-focused thematic discussions at VUF workshops and conferences, including, for example, Hai Phong's green growth. Sharing information on concrete urban development challenges in the city could provide an opportunity for all stakeholders, especially international donors, to plan activities in Hai Phong. The national government and international partners could consider extended collaboration with this initiative.

Box 4.8. **Viet Nam Urban Forum**

The Viet Nam Urban Forum (VUF) was set up in 2003 based on a consensus of government authorities and international and domestic organisations. It is led by Viet Nam's Ministry of Construction, which is also responsible for urban development in the country. Its mission is "to promote dialogue for sharing of experiences and knowledge among government agencies, associations, unions, scientific/professional, social/political organisations and non-governmental organisations, entities and individuals of all economic sectors and the donor agencies, with the aim of actively contributing to the formulation and implementation of policies on urban development and management, and improving the effectiveness of urban development and management activities in Viet Nam". The VUF works as a network and platform of discussion between these stakeholders, and is divided into seven member groups: multilateral donor organisations, bilateral donor organisations, government ministries and agencies, professional associations and academic institutions, cities, private sector corporations and NGOs. It has 150 members and plans to expand to 200 members by 2020.

The VUF organises three types of events and activities: i) annual events, consisting of the Annual Conference (General Assembly) and the Viet Nam Urban Day; ii) conferences, workshops and seminars based on thematic discussions; and iii) the production of policy recommendations. It provides an interesting opportunity to bring to the table local governments, the national government and international partners, to encourage urban green growth.

Source: Viet Nam Urban Forum (2012), *Viet Nam Urban Forum*, Version 6, May, www.geoviet.vn/FM/GetFile.aspx?ID=a2167cb2-8b36-4015-9af4-b68ca64d0db2 (accessed 18 June 2016).

Main policy recommendations

- **Review and revise national strategies and policies** in various sectors to encourage greater recognition of interdependence between central, provincial and local governments, and among Hai Phong and neighbouring cities and provinces, on their role in pursuing green growth objectives.
- Consider **a legal framework** to provide the basis for political, administrative and fiscal relations for each level of administration to draw up clear mandates for green growth in different sectors.
- Consider the use of a **National Strategic Reference Framework (NSRF)** as a basis for dialogue and co-ordination on green growth initiatives at the national level.
- Develop the concept of **functional urban areas** in Viet Nam and consider the creation of a **metropolitan planning institute**.
- Provide **funds** for multi-province green growth initiatives.
- Establish **a high-level forum** for co-ordination and collaboration, where senior members of government from the different cities can meet, discuss issues and agree on joint approaches and projects for Hai Phong, in open and consensus-based processes.
- Establish an interdepartmental **Green Growth Co-ordination Council** in the city that comprises a decision-making hierarchy and partnership for the implementation of the Local Action Plan on Green Growth, and/or form **a joint reviewing board for green growth** to review all green growth projects and ensure that they are aligned with projects in other departments.
- Amend national tax laws to improve **property taxation**, promote selective use of non-property taxes and user and benefit charges, and increase the flexible use of land rent beyond infrastructure investment.
- **Develop communication approaches** at both the ward and city levels, particularly in the areas of discussion and consultation on green growth. The city government could develop guidelines showing which kinds of information methods are appropriate for which materials.
- Make **official development finance** for the port contribute more comprehensively to green growth in Hai Phong, including its hinterland connections and its environmental performance.
- Set up a **local green investment fund** for urban development in Hai Phong.
- Expand **city-to-city collaboration** networks and the use of the **Viet Nam Urban Forum**.

Notes

1. For further details, see: Viet Nam Trade Promotion Agency (2014), "Achievements of Viet Nam '*Doi Moi*' Policy, www.vietrade.gov.vn/en/index.php?option=com_content&id=2256:achievements-of-viet-nam-doi-moi-policy-&Itemid=287 (accessed 20 June 2016).
2. See OECD (2015a) for further information.
3. See Roy and Langford (2008) for an in-depth discussion.
4. For further information on "Hai Phong Urban Transport Development" see: http://Haiphong.gov.vn/Portal/Detail.aspx?Organization=UBNDTP_En&MenuID=7270&ContentID=52671 (accessed 20 June 2016).
5. In Viet Nam, the proportion of fiscal revenue to GDP has increased, and since 2004, the percentage has risen to more than 25%. However, Viet Nam still suffers from a chronic fiscal deficit (Uchimura and Kono, 2012). This is attributable to three factors: the balance that is related to the carried-over funds, the reform of public servants' salaries, and the fiscal transfer to local governments (which accounts for more than half of the total amount).
6. For further information, see OECD (2014b), *Recommendation of the Council on Effective Public Investment Across Levels of Government*, adopted 12 March, 2014, www.oecd.org/regional/regional-policy/Principles-Public-Investment.pdf.
7. Given the centralised state apparatus in Viet Nam, urban governments tend to act as a lower tier of the central government (Huyen, 2014). Ha Noi, Hai Phong, Ho Chi Minh City, Da Nang and Can Tho are the five provincial cities that operate as the engines of socio-economic development. According to the classification of cities in Viet Nam, Ha Noi and Ho Chi Minh City are classified as special cities that serve as hubs to promote national development, whereas Hai Phong, Da Nang and Can Tho are classified as Class 1 cities, functioning as national and regional development centres. For further information see Government of Viet Nam (2009), *Decree No. 42/2009/ND-CP on the Grading of Urban Centers*, www.kenfoxlaw.com/resources/legal-documents/governmental-decrees/2342-vbpl-sp-2423.html (accessed 13 June 2016).
8. For further information on the guidelines, see OECD (2008b), *OECD Guiding Principles for Regulatory Quality and Performance*, http://dx.doi.org/10.1787/9789264056381-en.
9. For further information, see Government of Croatia, www.opengovpartnership.org/sites/default/files/Inspiring%20Story%20-%20Croatia.pdf.
10. For further information, see Australian Government: http://www.vic.gov.au/blog/research-reports/web-2-0-new-tools-democratic-conversations-snapshot-initiatives-government/(accessed 13 June 2016).

11. For further information, see The Danish Board of Technology Foundation: http://participedia.net/en/cases/board-technology-denmark (accessed 20 June 2016).

12. All the figures related to ODA given in this chapter are expressed in 2013 constant prices.

13. Bilateral ODA activities targeting environmental objectives can be identified in a dataset accessible through the OECD statistical website and the CRS database. This dataset contains commitment data on aid in support of environmental sustainability and aid targeting the objectives of the Rio Conventions (biodiversity, climate change mitigation, climate change adaptation and desertification). Multilateral ODA activities targeting environmental objectives can be identified in the CRS database accessible through the OECD statistical website. While non-concessional other official flows (OOF) are tracked in the OECD database, they are not marked against the objectives of the Rio Conventions. The analysis included in this report is thus limited, as it only focuses on ODA activities. Projects for urban areas were identified by a word search based on their purpose (e.g. "urban development and management"). Where the purpose field did not indicate whether the project was urban, individual project descriptions were examined. Similarly, the main element used to identify projects committed to rural areas was the classification of their purpose (i.e. "rural development"). If this was not conclusive, the titles and a short description of the projects was used to determine if they qualified. Agricultural, fishery and forestry projects were classified as rural. The remaining rural projects were identified based on their long description. Unspecified projects refer to all other projects, without detailed descriptions or details of their geographical scope.

14. Hai Phong also benefitted from projects financed by the International Development Association (IDA) of The World Bank, at USD 440 million. These projects also benefited Ho Chi Minh City, Can Tho and Nam Ding, without specifying how much of this amount was specifically spent on Hai Phong. They were therefore excluded from the calculations and integrated into the category "Other cities" in Figure 4.7.

15. The amounts committed for some of these grants are not available, and the actual amount is thus higher.

16. See: Climate Funds update, www.climatefundsupdate.org/data (accessed 16 July 2016).

17. See: Agence Française de Développement, www.afd.fr/home/pays/asie/geo-asie/afd-vietnam/cac-du-an-cua-afd-tai-viet-nam/phat-trien-o-thi/projets-d-appui-aux-fonds-urbains (accessed 18 June 2016).

18. See: Vietnam Development Bank, http://en.vdb.gov.vn/VDB/products--services/on-lending-oda-management (accessed 17 July 2016).

Bibliography

Abbott, J. (2012), *Collaborative Governance and Metropolitan Planning in South East Queensland – 1990 to 2010: From a voluntary to a statutory model*, ACELG, Queensland Government, Council of Mayors of South East Queensland, http://apo.org.au/resource/collaborative-governance-and-metropolitan-planning-south-east-queensland-1990-2010 (accessed 3 August 2016)

ADB (2012), *Assessment of Public-Private-Partnerships in Viet Nam, Constraints and Opportunities*, Mandaluyong City, Philippines.

Agence Française de Développement (2015), *Urban Development Projects*, www.afd.fr/home/pays/asie/geo-asie/afd-vietnam/cac-du-an-cua-afd-tai-viet-nam/phat-trien-o-thi/projets-d-appui-aux-fonds-urbains (accessed 18 June 2016)

Asian Exim Banks website, http://www.asianeximbanks.org/content/institutions/vietnam-development-bank (accessed 01 August 2016)

Bahl, R., J. Linn and D. Wetzel (2013), "Governing and financing metropolitan areas in the developing world", in R. Bahl, J. Linn and D. Wetzel (eds.) (2013), *Financing Metropolitan Areas in the Developing World*, Lincoln Institute of Land Policy, Cambridge, Massachusetts. pp 1-30.

Campbell, T. (1998), "City development options for Hai Phong – Charting a path to the year 2020", Urban Partnership TWU, www.ucl.ac.uk/dpu-projects/drivers_urb_change/urb_economy/pdf_urban_dev_finance/Urban_Partnership_TWU_Campbell_City_Dvlpt.pdf.

CECODES, VFF-CRT & UNDP (2015), *The Viet Nam Provincial Governance and Public Administration Performance Index (PAPI) 2014: Measuring Citizens' Experiences*, joint Policy Research Paper by Centre for Community Support and Development Studies (CECODES), Centre for Research and Training of the Viet Nam Fatherland Front (VFF-CRT) and the United Nations Development Programme (UNDP), Ha Noi, Viet Nam, http://papi.vn/documents/provincial-profiles/2014/en/PAPI2014_FinalReport_SmallSize_ENG.pdf.

City of Hai Phong (2015), "Answers to the OECD case study questionnaire", internal document, unpublished.

Clean Air Asia (2012), "Accessing Asia; air pollution and greenhouse gas emissions indicators for road transport and electricity". http://cleanairasia.org/wp-content/uploads/portal/files/Accessing_Asia_2012_Edition_Guidelines.pdf.

Climate Funds update (n.d.), *The Latest Information on Climate Funds*, www.climatefundsupdate.org/data (accessed 16 July 2016).

Darko, E., N. Manh Hai and S. Whitley (2015), "Mapping current incentives and investment in Viet Nam's transport sector: Informing private climate finance", Overseas Development Institute, London, https://www.odi.org/sites/odi.org.uk/files/odi-assets/publications-opinion-files/9642.pdf.

Dudek, C.M. (2005), "Co-ordinating regional policy in the EU", Department of Political Science, Hofstra University, http://aei.pitt.edu/3170/1/Regional_Policy_Paper-Dudek1-1.pdf.

Ellis, P. and M. Roberts (2016), "Leveraging urbanisation in South Asia: Managing Spatial Transformation for Prosperity and Livability", World Bank Group, Washington DC, https://openknowledge.worldbank.org/handle/10986/22549 (accessed July 2016).

Friends of Midcoast Maine (n.d.), *Friends of Midcoast Maine – A grassroots initiative for sensible growth*, https://lcrpc.org/uploads/visual_edit/preservingsmalltowncharacter-1.pdf (accessed 9 July 2016).

Fritzen, S. (2006), "Probing system limits: Decentralisation and local political accountability in Viet Nam", *Asia-Pacific Journal of Public Administration*, Vol. 28(1), pp.1-24.

General Statistics Office (2015), *Hai Phong Statistics Yearbook*, Ha Noi, Viet Nam.

Government of France (2015), *Schéma de coherence territoriale (SCoT), Ministère du Logement, de l'Égalité des Territoires et de la Ruralité*, http://www.logement.gouv.fr/schema-de-coherence-territoriale-scot (accessed 3 August 2016).

Government of Viet Nam (2014), *National Action Plan on Green Growth for Period 2014-2020*, Socialist Republic of Viet Nam, Ha Noi. https://www.giz.de/en/downloads/Decision_403-2014-TTg_EN.pdf.

Government of Viet Nam (2012a), *Viet Nam National Green Growth Strategy for the Period 2011-2020, with a Vision to 2050*, Socialist Republic of Viet Nam, Ha Noi, https://www.giz.de/en/downloads/VietNam-GreenGrowth-Strategy.pdf.

Government of Viet Nam (2012b), *Viet Nam Sustainable Development Strategy for 2011-2020*, Socialist Republic of Viet Nam, Ha Noi, https://sustainabledevelopment.un.org/content/documents/995vietnam.pdf.

Government of Viet Nam (2011a), *Viet Nam's Socio-economic Development Strategy for the Period of 2011-2020* [unofficial translation by www.economica.vn], Socialist Republic of Viet Nam, Ha Noi.

Government of Viet Nam (2011b), *National Strategy on Climate Change*, Socialist Republic of Viet Nam, Ha Noi, www.chinhphu.vn/portal/page/portal/chinhphu/trangchu (accessed August 2016).

Government of Viet Nam (2009a), *Law on Urban Planning*, Socialist Republic of Viet Nam, Ha Noi, http://moj.gov.vn/vbpq/en/lists/vn%20bn%20php%20lut/view_detail.aspx?itemid=10461 (accessed August 2016).

Government of Viet Nam (2009b), *Decree No. 42/2009/ND-CP on the Grading of Urban Centers*, www.kenfoxlaw.com/resources/legal-documents/governmental-decrees/2342-vbpl-sp-2423.html (accessed 13 June 2016).

Huyen, D. (2014), "A Tale of Five Cities: How centrally governed municipalities fair in citizens' assessment of governance performance", *Viet Nam Law and Legal Forum*, UNDP Viet Nam, July-August, pp 9-14, 47, www.academia.edu/8039106/A_Tale_of_Five_Cities_How_centrally_governed_municipalities_fair_in_citizens_assessment_of_governance_performance (accessed 5 July 2016).

Koh, D., D. Du Dam and T.K.C. Nguyen (2009), "Government organization structure and excellent public services: The case of Viet Nam and some recommendations for change", in UNDP *Reforming Public Administration in Viet Nam, Current situation and Recommendations*, the National Political Publishing House, http://www.vn.undp.org/content/vietnam/en/home/library/democratic_governance/reforming-public-administration-in-viet-nam--current-situation-a.html (accessed July 2016).

Konrad Adenauer Foundation and Association of Cities of Viet Nam (2009), "People's participation and local governance: Challenges and opportunities for municipalities", European Union, Konrad Adenauer Foundation, the Association of Cities of Viet Nam, www.kas.de/wf/doc/kas_17873-1522-2-30.pdf?091019054634 (accessed 16 July 2016).

Laegrid, P., A. Dyrnes Nordo and L.H. Rykkja (2013), "The quality of co-ordination in the Norwegian central government: The importance of co-ordination arrangements and structural, cultural and demographic factors", *COCOPS Working Paper No. 14*, European Commission, European Research Area, www.cocops.eu/wp-content/uploads/2013/12/Working-Paper-no.-14.pdf.

Matsumoto, T. and L. Daudey (2014), "Urban Green Growth in Dynamic Asia: A Conceptual Framework", OECD Regional Development Working Papers, No. 2014/12, OECD Publishing, Paris. DOI: http://dx.doi.org/10.1787/5js7svlw8m0x-en.

Matsumoto, T. and C. Nuttall (2014), "Integrating subnational action", in Green Growth Best Practice, *Green Growth in Practice: Lessons from Country Experiences*. Available at: http://www.ggbp.org/report/best-practice-report/integrating-subnational-action (accessed 20 August 2016).

Nguyen, T. et al. (2015), "Citizen participation in city governance: Experiences from Viet Nam", *Public Administration and Dev*elopment, John Wiley and Sons, No. 35, pp. 34-45, http://onlinelibrary.wiley.com/doi/10.1002/pad.1702/pdf.

Nguyen, T. et al. (2013), *Municipal Government in Viet Nam: Case studies of Ho Chi Minh City and Da Nang*, University of Economics and Law, the Asia Foundation, https://asiafoundation.org/resources/pdfs/MunicipalGovermentinVN.pdf.

OECD (forthcoming), *Green Growth in Bandung, Indonesia*, OECD Publishing, Paris.

OECD (2016), "Creditor Reporting System: Aid activities targeting Global Environmental Objectives", *OECD International Development Statistics* (database), http://dx.doi.org/10.1787/9c778247-en.

OECD (2015a), *OECD Territorial Reviews: Valle de México, Mexico*, OECD Publishing, Paris, http://dx.doi.org/10.1787/9789264245174-en.

OECD (2015b), *OECD Urban Policy Reviews: China 2015*, OECD Publishing, Paris, http://dx.doi.org/10.1787/9789264230040-en.

OECD (2015c), *OECD Urban Policy Reviews: China 2015*, OECD Publishing, Paris, http://dx.doi.org/10.1787/9789264230040-en.

OECD (2015d), *Green Growth in Bangkok, Thailand*, OECD Publishing, Paris, http://dx.doi.org/10.1787/9789264237087-en.

OECD (2015e), *Subnational Governments in OECD Countries: Key data* (brochure), www.oecd.org/gov/regional-policy/Subnational-governments-in-OECD-Countries-Key-Data-2015.pdf.

OECD (2014b), *Recommendation of the Council on Effective Public Investment Across Levels of Government*, adopted 12 March, 2014, www.oecd.org/regional/regional-policy/Principles-Public-Investment.pdf.

OECD (2013a), *Fiscal Federalism 2014: Making Decentralisation Work*, OECD Publishing, Paris, http://dx.doi.org/10.1787/9789264204577-en.

OECD (2013b), Green Growth in Kitakyushu, Japan, OECD Publishing, Paris, http://dx.doi.org/10.1787/9789264195134-en.

OECD (2013c), *Public Procurement Review of the Mexican Institute of Social Security: Enhancing Efficiency and Integrity for Better Health Care*, OECD Public Governance Reviews, OECD Publishing, Paris, http://dx.doi.org/10.1787/9789264197480-en.

OECD (2011), *Public Servants as Partners for Growth: Toward a Stronger, Leaner and More Equitable Workforce,* OECD Publishing, Paris, http://dx.doi.org/10.1787/9789264166707-en.

OECD (2010b), *OECD Territorial Reviews: Sweden 2010*, OECD Publishing, Paris, http://dx.doi.org/10.1787/9789264081888-en.

OECD (2010a), "Managing competencies in government: State of the art practices and issues at stake for the future", GOV/PGC/PEM(2010)1, OECD, Paris. http://www.oecd.org/officialdocuments/publicdisplaydocumentpdf/?cote=GOV/PGC/PEM(2010)1/FINAL&docLanguage=En.

OECD (2008a), *The State of the Public Service*, OECD Publishing, Paris, http://dx.doi.org/10.1787/9789264047990-en.

OECD (2008b), *OECD Guiding Principles for Regulatory Quality and Performance*, OECD Publishing, Paris, http://dx.doi.org/10.1787/9789264056381-en.

OECD (2005), *Modernising Government: The Way Forward*, OECD Publishing, Paris, http://dx.doi.org/10.1787/9789264010505-en.

OECD (2001), *Citizens as Partners: OECD Handbook on Information, Consultation and Public Participation in Policy-Making*, OECD Publishing, Paris, http://dx.doi.org/10.1787/9789264195578-en.

OECD/Korea Institute of Public Finance (2012), *Institutional and Financial Relations across Levels of Government,* OECD Fiscal Federalism Studies, OECD Publishing, Paris, http://dx.doi.org/10.1787/9789264167001-en.

OECD (1998), *Recommendation of the Council on Improving Ethical Conduct in the Public Service Including Principles for Managing Ethics in the Public Service*, C(98)70/FINAL, OECD, Paris (unpublished material).

Oliver, J.E. (2000), "City size and civic involvement in metropolitan America", *American Political Science Review*, American Political Science Association, Vol. 94(2), pp. 361-373, www.jstor.org/stable/2586017?seq=1#page_scan_tab_contents (accessed 16 June 2016).

People's Committee of Hai Phong (2010), *Governance and Transparency Action Plan*, Department of Transport, http://pmuhaiphong.gov.vn/Du_an_PTGTDT_HP/22.Dec.10_HPUTDP_GTAP_EN.pdf.

Roy, J. and J. Langford (2008), "Integrating service delivery across levels of government: Case studies of Canada and other countries", *Collaboration: Networks and Partnerships Series*, IBM Center for the Business of Government, www.businessofgovernment.org/report/integrating-service-delivery-across-levels-government-case-studies-canada-and-other-countries (accessed 16 July 2016).

Savery N. and L. de Chastel (2009), "The importance and value of urban and regional planning", *Australian Planner*, Vol. 16(3), pp. 6-9, http://dx/doi.org/10.1080/07293682.2009.10753400.

Tanzi, V. and H. Davoodi (1997), "Corruption, public investment, and growth", *IMF Working Papers*, WP/97/139, International Monetary Fund, Washington DC, www.imf.org/external/pubs/ft/wp/wp97139.pdf.

Uchimura, H. and H. Kono (2012), "Intergovernmental fiscal relationships in Viet Nam after the 2002 State Budget Law: The Centre-Province and the Province-District/Commune Dimensions" in H. Uchimura (ed.), *Fiscal Decentralization and Development: Experiences of Three Developing Countries in Southeast Asia*, IDE-JETRO, Palgrave MacMillan, pp. 101-131, http://www.ide.go.jp/English/Publish/Books/Palgrave/b020.html (accessed July 2016).

UNDP (2009), *Reforming Public Administration in Viet Nam, Current situation and Recommendations*, the National Political Publishing House, http://www.vn.undp.org/content/vietnam/en/home/library/democratic_governance/reforming-public-administration-in-viet-nam--current-situation-a.html (accessed July 2016).

van Arkadie, B. and R. Mallon (2003), *Viet Nam: A Transition Tiger? The Introduction of Doi Moi*, Asia Pacific Press at The Australian National University.

Vietnam Development Bank (n.d.), *On-lending ODA management*, http://en.vdb.gov.vn/VDB/products--services/on-lending-oda-management (accessed 17 July 2016).

Viet Nam Trade Promotion Agency (2014), "Achievements of Viet Nam *'Doi Moi'* Policy", 11 December, www.vietrade.gov.vn/en/index.php?option=com_content&id=2256:achievements-of-viet-nam-doi-moi-policy-&Itemid=287 (accessed 20 June 2016).

Viet Nam Urban Forum (2012), *Viet Nam Urban Forum*, Version 6, May, www.geoviet.vn/FM/GetFile.aspx?ID=a2167cb2-8b36-4015-9af4-b68ca64d0db2 (accessed 18 June 2016).

World Bank (2014a), *Revising Viet Nam's State Budget Law (2002): Proposals Drawing on International Experience,* policy note, World Bank, UKAid, Australian Aid, www.fetp.edu.vn/attachment.aspx?ID=34407 (accessed 2 July 2016).

World Bank (2014b), "Viet Nam Development Report 2014 – Skilling up Viet Nam: Preparing the workforce for a modern market economy – overview report", World Bank, www-wds.worldbank.org/external/default/WDSContentServer/WDSP/IB/2013/11/26/000461832_20131126115026/Rendered/PDF/829400AR0ENGLI0Box0379879B00PUBLIC0.pdf.

World Bank (2011), *Viet Nam Urbanization Review*, World Bank, https://openknowledge.worldbank.org/bitstream/handle/10986/2826/669160ESW0P1130Review000Full0report.pdf.

World Bank (2010), *Viet Nam Development Report 2010 – Modern Institutions*, World Bank, www-wds.worldbank.org/external/default/WDSContentServer/WDSP/IB/2010/03/19/000333037_20100319004055/Rendered/PDF/535730WP0P11221ng0reduced0file0size.pdf.

World Economic Forum (2013a), *The Green Investment Report: The Ways and Measures to Unlock Private Finance for Green Growth,* Green Growth Action Alliance, Geneva, http://www3.weforum.org/docs/WEF_GreenInvestment_Report_2013.pdf.

World Economic Forum (2013b), *Financing Green Growth in a Resource-constrained World: Partnerships for Triggering Private Finance at Scale*, Geneva, http://www3.weforum.org/docs/WEF_ENI_FinancingGreenGrowthResourceConstrainedWorld_Report_2012.pdf.

World Economic Forum (2013b), *Viet Nam Country Profile*, http://www3.weforum.org/docs/GCR2014-15/Viet Nam.pdf.

ORGANISATION FOR ECONOMIC CO-OPERATION AND DEVELOPMENT

The OECD is a unique forum where governments work together to address the economic, social and environmental challenges of globalisation. The OECD is also at the forefront of efforts to understand and to help governments respond to new developments and concerns, such as corporate governance, the information economy and the challenges of an ageing population. The Organisation provides a setting where governments can compare policy experiences, seek answers to common problems, identify good practice and work to co-ordinate domestic and international policies.

The OECD member countries are: Australia, Austria, Belgium, Canada, Chile, the Czech Republic, Denmark, Estonia, Finland, France, Germany, Greece, Hungary, Iceland, Ireland, Israel, Italy, Japan, Korea, Latvia, Luxembourg, Mexico, the Netherlands, New Zealand, Norway, Poland, Portugal, the Slovak Republic, Slovenia, Spain, Sweden, Switzerland, Turkey, the United Kingdom and the United States. The European Union takes part in the work of the OECD.

OECD Publishing disseminates widely the results of the Organisation's statistics gathering and research on economic, social and environmental issues, as well as the conventions, guidelines and standards agreed by its members.

OECD PUBLISHING, 2, rue André-Pascal, 75775 PARIS CEDEX 16
(04 2016 07 1 P) ISBN 978-92-64-26016-0 – 2016

www.ingramcontent.com/pod-product-compliance
Ingram Content Group UK Ltd.
Pitfield, Milton Keynes, MK11 3LW, UK
UKHW051503290926
15471UKWH00008B/1389

9 789264 260160